W9-BXN-342

NEWTON'S GIFT

How Sir Isaac Newton Unlocked the
System of the World

DAVID
BERLINSKI

The Free Press

NEW YORK   LONDON   TORONTO
SYDNEY   SINGAPORE

$f\mathbf{P}$

THE FREE PRESS
A Division of Simon & Schuster Inc.
1230 Avenue of the Americas
New York, NY 10020

THE FREE PRESS and colophon are trademarks
of Simon & Schuster Inc.

Designed by Kyoko Watanabe

Manufactured in the United States of America

10   9   8   7   6   5   4   3   2   1

Library of Congress Cataloging-in-Publication Data

Berlinski, David
Newton's gift : how Sir Isaac Newton unlocked the system of the world /
David Berlinski.
p.   cm.
Includes index.
1. Newton, Isaac, Sir, 1642-1727.   2. Physics--History.
3. Physicists—Great Britain—Biography.   I. Title.
QC16.N7 B48 2000
530'.092—dc21
[B]   00-34724

ISBN 0-684-84392-7

*Dedicated to
my parents at ninety*

CONTENTS

CONTENTS

They reckon ill that leave me out,
With me they fly, I am the wings,
I am the doubter and the doubt,
And I the hymn the Brahmin sings.

INTRODUCTION

Isaac Newton is the largest figure in the history of western science, his influence both inescapable and immeasurable. Newton created the disciplines of rational and celestial mechanics; he discovered the calculus; he advanced a theory of color; and he made profound and audacious contributions to pure mathematics, optics, and astronomy. By showing that a mathematical investigation of the physical world was possible, he made that investigation inevitable.

Newtonian mechanics is not only the first, but the greatest, of scientific theories. It provides an explanation for a wide range of terrestrial and celestial phenomena. Within its proper domain of application, it is extraordinarily accurate. And it embodies a combination of simplicity and scope still denied any other scientific theory.

These are very considerable virtues. They explain some but not all of Newton's influence.

Newton's masterpiece is the *Mathematical Principles of Natural Philosophy*, or the *Principia*, as it is generally called (after its Latin title). Nothing like the *Principia* had ever appeared before the seventeenth century; and in truth, nothing like the *Principia* has ever appeared afterward. In very large measure, it was the *Principia* that ignited the furious dark energies that brought mathematical physics into existence and that have sustained its fires for more than three hundred years.

The Newtonian universe is mechanical in the sense that like a clock it is self-sustaining. There is order everywhere. Planets proceed sedately along their appointed paths, holding themselves in a state of equipoise. Physical processes take place within an unchanging vault of absolute space and in accord with the unchanging beat of absolute time. Propelling itself through space, the universal force of gravitation subordinates all material objects to a single modality of attraction. And all this proceeds in accordance with simple mathematical laws.

Newton's great vision of what he called *the system of the world* has set the agenda for research for more than three hundred years. As the twenty-first century commences, physicists are searching for the unified theory that by means of one set of unutterably pregnant laws would explain the properties of matter in all of its manifestations. The terms of the search are by now familiar. But they are Newton's terms and before Newton, the search would have made little sense. With the theory complete, physics will have reached its appointed end simply because it has no place further to go. Everything will have been understood. Science as an intellectual activity will continue to amass facts in biology or chemistry or psychology, but those facts are destined to be amassed within the chambers of a cathedral that has already been completed. A deep silence will prevail.

If Newton's *Principia* has given the future of mathematical physics its characteristic shape, it has given the future its characteristic question as well. The Newtonian universe is a closed physical system. Whatever happens takes place as the result of causal interactions between material objects. There is nonetheless one aspect of the Newtonian world that is not explained by Newton's theory, and that is Newton's theory itself. The law of universal gravitation binds the world's farflung particles into a coherent whole; but the law is itself transcendent. It cannot be given an explanation in material terms.

This is true as well for the equations governing the electromagnetic field, Einstein's field equations for general relativity, and Schrödinger's wave equation in quantum mechanics. The laws of nature by which nature is explained are not themselves a part of nature. No physical theory predicts their existence nor explains their power. They exist beyond space and time; they gain purchase by an act of the imagination and not observation, they are the tantalizing traces in matter of an intelligence that has so far hidden itself in symbols. Efforts to explain the laws of nature in terms of still further laws of nature that explain themselves have been unavailing. They are what they are.

The great physicists have always recognized that the organization of nature represents a profound mystery. They have for this reason paid homage to those laws, seeing in their symmetry and perfection something of great and ineffable majesty.

In the utterance of this sentiment, they are following in Newton's broad wake, paying homage to what he paid homage to, a captive in the end of his command.

A NOTE TO THE READER

I have made no effort in this little book to explain the whole of Newton's scientific achievement, and in particular I do not discuss Newton's second masterpiece, *The Opticks*, at all. *Ars longa vita brevis*. Neither have I attempted to write another version of Isaac Newton's life. There are many fine contemporary biographies available in all of the European languages. Richard Westphall's *Never at Rest* is widely considered the standard account. My aim instead has been to offer a sense of the man without specifying in detail his day-to-day activities. This is much harder than it seems. Newton lived within himself. He rarely traveled and never traveled beyond England. He kept few friends and his correspondence is largely scientific. Newton wrote at great length, but he almost never wrote about himself in the candid fashion of contemporary scientific figures. He was conscious of his powers, of course, and justifiably proud of his achievements, but beyond attributing them to his own matchless capacity for hard work, he did not attribute them to anything at all. Despite the brilliance of his reflected light, he himself stands where I suspect he wished to stand—in the shadows.

Nothing in this book requires very difficult mathematics.

A Note to the Reader

I have tried to illustrate every mathematical idea in three different ways: first in words, second in symbols, and third by means of pictures or diagrams. Newton's ideas and the mathematical language that he used are unfamiliar, but they can be grasped in terms of mental motions in which some things tend toward other things, processes converge toward distant limits, and planets are represented as points in the night sky. This is, in fact, how Newton grasped them.

⊰ 1 ⊱

IN THE YEAR THAT
GALILEO DIED

I SAAC NEWTON WAS BORN IN THE HAMLET OF Wolingsthorpe in 1642 and died in London in 1727. His final hours were unrelieved. With their usual masterly incompetence, his physicians, Richard Mead and William Cheselton, presumed that Newton was suffering from bladder stones, a diagnosis that accommodated his incontinence but shed little light on his continual coughing. They had examined their patient diligently in his narrow bedroom at Kensington, obtaining what physicians inevitably obtain on palpating a patient, namely a sonorous grunt and an admission of pain, and after plumping his pillows and pointlessly radiating their own good health into the close bedchamber, they confessed to Newton's devoted nephew by marriage, John Conduitt, what was in any case evident, that their patient was dying. No further medical assistance was likely to be of use. Newton was in great pain, perspiration

1

forming on his brow and then drenching his face as he was attacked by the spasms characteristic of his disease, but when
they relaxed their grip, his mind reacquired its habitual lu-
cidity and he was able to talk with visitors who had assembled
to pay their respects, inquiring after their health and wishing
them well as with a great cold bustle they vacated his melan-
choly bedroom. The clock in the downstairs hallway continued to mark the passing of the minutes, then the hours, and
then the days. What was destined to become became. Mention
was at last made of the Sacraments. He would not, Newton
told his nephew and niece, accept the last rites of the church.
Newton rallied briefly and then declined, dying with enormous energy early in the morning of the 20th of March. The
Royal Society, over which he had presided with immense dig-
nity, found precisely the lapidary words to commemorate his
passing: "The Chair being vacant by the death of Sir Isaac
Newton, there was no meeting this day."

Like Eve, the Isaac Newton of myth and memory—*our*
Newton, of course—was prompted by an apple, seeing sometime in the spring of 1662 that apples when falling fall
downward. His biographer William Stukeley respectfully
conveyed the essence of the myth: "After dinner," he re-
counts, "the weather being warm, we went into the garden
and drank tea, under the shade of some apple trees, only he
and myself.

> Amidst other discourse, he told me, he was just in the
> same situation, as when formerly, the notion of gravita-
> tion came into his mind. It was occasioned by the fall of
> an apple as he sat in a contemplative mood. Why should
> that apple always descend perpendicularly to the ground,

thought he to himself. Why should it not go sideways or upwards, but constantly to the earth's center?

I have always understood the apple to have fallen on Newton's *head* with an invigorating *boink;* but I may have been misinformed. Stukeley clearly has Newton looking at the apple as it fell; but the charming thing about the story, which in time traveled as far afield as Japan, is that like all good stories, it seems to contain a mystery at the core of its narrative marrow, the falling apple followed by a thought-inducing *boink* still retaining all of its old and troubling suggestiveness. An apple? Falling? *Yes,* but why *downward?*

If they gave the matter any thought at all, ancient writers argued that falling objects fall downward because the earth is their natural resting place, this observation persisting as received wisdom if only because it seemed and still seems self-evident. Apples have been falling since time immemorial and in the end every last one of them has fallen downward. If the concept of a natural resting place seems suspiciously to suggest that the apples have a voice in the scheme of things and would all in all prefer to lie on the ground, then it is easy enough to rehabilitate received wisdom by means of a few minor renovations. *All* material objects fall toward the center of the earth. An apple is a material object. What more can one say?

What Newton in fact did say, Stukeley remembered, is that "there must be a drawing power in matter," the aforementioned thought-inducing *boink* prompting Newton to a series of reflections progressively at odds not only with common sense but with the common world that common sense reflects. At first, it is true, the notion of a "drawing power" in material objects suggests a physical question answered by a metaphysical evasion, a drawing power kin to the notorious dormitive power by which schoolmen explained the fact that opium induces sleep. Like the letter "c," question and answer describe a shape in prospect of becoming a circle. Traveling around its backside, we ask why opium induces sleep. Ascending smoothly its flank, we learn that opium has a dormitive power. A spark now shuttles between the typesetter's gap. A dormitive power is just the power to induce sleep, so that by the time one revolution of the circle has been completed, we have understood only that a substance with the power to induce sleep has the power to induce sleep.

Yet appearances notwithstanding, Newton's reflections are different; they are at once radical and profound. "If matter thus draws matter," Stukeley remembered Newton remembering, "it must be in proportion to its quantity." And it is here, within the compass of eight words, that Newton hops off the letter "c," making for territory no man had claimed before. To be sure, the elegant asseveration that matter thus draws matter reminds us of what is antecedently known: things fall downward; but the words that follow draw a *mathematical* connection between a qualitative power and a quantitative property. Before the connection is made, the drawing power is identified only by the circumstance that it is capable of doing some drawing; but afterwards, everything changes, that drawing power now connected by proportionality to the quantity of an object, what later physicists would call its mass. The greater the mass, the greater the drawing power. The halter of specificity has been imposed on heretofore disorderly concepts. The power in matter to draw matter has for the first time been given a quantitative expression.

The inference continues: it gathers force. If matter does draw matter, Stukeley goes on to say (writing on behalf of a Newton presumably nodding in agreement), "therefore the apple draws the earth as well as the earth draws the apple." We are now fourteen words beyond the obvious. The idea that the fall of an apple toward the center of the earth might be an example of a reciprocal and reciprocating relationship between the earth and the apple, with the earth tugging at the apple and the apple tugging at the earth in return—this represents a genuine novelty in the history of thought. And for obvious reasons. The earth is large, the apple small; but Newton realized, apparently all at once, that if the drawing power of matter is proportional to its quantity, *every* physical object must logically have some drawing power of its own.

The inference now shoots out like light shining from shook foil. There is the real world in which apples fall downward; there is the revealed world, in which material objects attract one another, revelation turning on the mathematical hinge that material objects attract one another in proportion to their quantity. But what holds of the green earth and its golden apples holds in virtue of the fact that both are material objects and it holds only in virtue of the fact that both are material objects. Their shape is irrelevant; so, too, their color, consistency, tang, hue, friability, texture, or beauty. It follows that ". . . there is a power," Stukeley concluded, "like that we here call gravity, which extends itself throughout the universe." A series of local observations—this apple, that tree, an orchard—has propelled itself throughout the whole of space, binding every material object to every other material object.

Some apple. Some story.

❁   ❁   ❁

ISAAC NEWTON was born during the year that Galileo died; and what is more, his own father died while he was yet unborn, so that when finally he emerged on the scene, a somewhat weak and sickly infant, Newton had already lost both his intellectual and his biological fathers, a member in good standing of the worldwide fraternity of fatherless sons. Surviving against the expectations of his nurse, Newton was baptized the following year, the parish priest placing an oiled finger on his forehead with great solemnity but little conviction; and there follows after that (because there must have followed after that) a sunny splotch of muted warmth when like all infants Newton regarded himself as a precarious extension of his mother's body and lived entirely within the sphere of her smells, her skin, and the melodies of her voice. In 1646, Newton's mother, Hannah Ayschoughs, married the Reverend Barnabas Smith, who has entered history as a cloudburst of cold emotions, for with an alacrity that today seems unseemly, Hannah vacated Woolsthorpe Manor to live some miles away with the elderly Smith, leaving a barely sentient Newton to be raised by his maternal grandparents. The effect must have been one of a blow delivered against a wound. Every society regards its children through a telescope of sorts, the focusing power of the instrument changing with each generation, but however Newton's grandparents regarded Newton, it is certain only that they regarded him with very little sentimentality. It is possible to imagine the scene: Newton's mother abruptly informing the barely comprehending child of her departure, this followed by his agonized sharp shocking realization that *she* meant to leave *him*. Enter the grandparents, familiar figures, of course, but in their fifties already ancient, evil smelling with age, hobbled, coarse, and above all cold.

"*Ay*, he'll be just fine now, Hannah, you leave that to us."

Newton survived the blow, but the wound was deep and his personality acquired something burrowing in its nature, a desire for withdrawal persisting until middle age, some effort to get under the earth and seal up the apertures to his soul, like a rabbit trembling in the last of its lairs, listening always for time's badger. When in 1653 his hideously priapic stepfather obligingly died after fathering two additional children, his mother returned from her strange exile; having seen two husbands decamp for the country churchyard, she was now wealthier than when she had departed, and while her reappearance cannot have been the cause of dissatisfaction at Woolsthorpe, there is no record anywhere of Newton having dissolved the carapace of his personality in an efflorescence of tenderness. He remained throughout his long life devoted to his mother or to her memory, but he had been wounded and he did not forget.

There follows the detritus of childhood, the tang and texture unrecoverable after three hundred years, with only a few anecdotes poking their way up from the otherwise stony soil. Having been reunited with his mother at the age of ten, Newton was sent two years later to the King Edward Grammar School in Grantham, some eight kilometers from Woolsthorpe (and so, far from home), where, like every other student, he was drilled in Latin, some Greek, and the Old and New Testaments. The classical languages he studied by the immemorial method of rote repetition. It is the only way, I suspect, to learn a classical language, and it is surely the only way to learn one well.

It is unsettling nonetheless to think of Newton being *drilled* in anything. There is the usual babble of distracted boys, their great thick plowman's fingers grasping uneasily at the chalk, the master tapping his rod ominously against his thigh. Newton is staring into space. The master asks him to decline a Latin verb. Newton turns his head from the small window and proceeds, his unmusical voice sounding in the little classroom as if it had come from a great distance.

Whatever his pedagogical talents, the school's master, Dr. Henry Stokes, had early on a fine appreciation of Newton's intellectual power and ultimately persuaded his mother against her better judgment to send him to university at Cambridge. No doubt, he expected that at least some of the glory of his protégé would redound to the benefit of the school, with he himself warming his hands in the cheerful warmth of his own perspicuity. In one of the ironies of history, Dr. Stokes remains fixed in the seventeenth century only as the man who first noticed Newton's genius, the conflagration that he engendered hot enough to leave little more than the evidence of a match in his scorched fingertips.

Let me see. There is not much here in the barrel of his early life. The classroom was cold and often damp and smelt deeply of rosin and chalk dust. Like most students, Newton boarded away from home, living in an apothecary's shop in Grantham. At some point, Newton's classmates came to appreciate Newton's frightening mental powers; he was not only odd, they realized, but incomprehensibly odd. Biographers and contemporaries alike recall a fight. Provoked by a blow, Newton challenged his antagonist after class and thrashed him. I am searching the bottom of the barrel now, winking tidbits into the light of day.

On one thing, all commentators do agree: as a very young man, Newton had an extraordinary mechanical gift, one that enabled him to construct ingenious models, make working windmills in miniature, see into the heart of machinery and hear the secret beat of wheels and cogs and meshed chains. It was more, this gift, than any kind of handiness or even dexterity; something in Newton's soul could sense and then seize the flow of energy and work in things, and where the sturdy men hauling grain to be milled saw only flat and somber sails turning in the English wind—there was only one working windmill in Grantham—Newton could discern an abstract alembic of impressed forces and movements, so that by the time he had studied some mechanical contraption, he knew not only how it was made, but the principles by which it worked, and with his astonishing capacity to move easily between scales of size, he was able to reconstruct in miniature a perfectly functioning windmill, complete with tower, grinding stones, conveyances for grain, and delicate translucent sails.

The seasons of the English countryside appear and reappear and for the first and only time, there is the faintest flushed suggestion of an infatuation in Newton's life, this with the apothecary's stepdaughter, Catherine Storer. Commentators and biographers are reticent about this relationship, but the common experience of men everywhere suggests the immense burden of unfulfilled tenderness that must have been conveyed by the chaste adhesion of her pink fingertips in his pale hand. Whatever the romantic stirrings prompted by these sentiments, Newton failed to act upon them, and it was left to Catherine Storer, almost seventy years later, when Newton had already been promoted to the Pantheon and

stood looming over English intellectual life, to look back on
the unrecoverable past and with a sigh no doubt designed to
conflate memory and myth, reveal to Newton's biographers
that once when she was young she had held the great man's
heart within her hands.

⫷ 2 ⫸

AN ESCAPE FROM
THE PLOW

FOR A TIME DURING HIS ADOLESCENCE, HIS
mother thought that Newton might come into
possession of Woolsthorpe Manor; she envi-
sioned, no doubt, the farm spreading further into the future,
her son and heir prudently managing the livestock and culti-
vating the crops, living as she had done by the season's sedate
clock, his wife a local village girl thrilled to have acquired
both a husband in Newton and a prospect in the farm, stout
and pregnant with a trailing line of Newtons. If nothing else,
this fantasy was calculated to complete the promotion of
Newton's side of the family from its laboring origins to a secure position in the gentry, his name, and so his mother linked
to the land by right of ownership, title, and long possession.
There followed a period of comical inadvertence, as the gan-
gling Newton, a book in hand and a vaguely irritated expression on his otherwise unlined face, is sent off to herd swine or

tend sheep or mend fences or otherwise place himself in the service of the rural menial. That maternal scheme now shrivels. The sheep trotting behind him and bleating in the summer air, Newton makes for the far pasture and reposes himself beneath a tree; the sheep spread out, flock and master both delighting in the dawning sense that this far from the manor house, no one expects anything of either. Seated peaceably, Newton reads, entirely unaware of the sheep, which have by now managed to wander off into a neighbor's field, where they amiably defecate on newly seeded hay and crop the luscious grass meant for other sheep. The sun rises to its apogee and then declines, Newton remaining lost in thought. Some time later, he reappears at the manor house; half the flock is yet enjoying a late afternoon carouse in a neighbor's field and the sheepdog, his responsibilities abrogated, has wandered off to pursue an interesting if feculent scent by the stream running through the meadow. Newton stands indolently in front of his mother, who regards him with a hard cold practical eye. She sees her family's ruin in his admission that he could not do what he did not want to do.

She must have brought her complaints to the men she knew. When Humphrey Babbington, Mrs. Clark's brother—she of the apothecary's shop—or Henry Stokes, or even her own brother William, pleaded with her to allow Newton to attend the university, she listened, half-inclined to argue that *that* one—there followed a jerk of her head toward the upstairs room where Newton is again doing nothing more and nothing better than reading—needs to be taken down a peg; but in the end, after hearing out stories of her son's unusual abilities and listening patiently to homilies in which rivers find their own true course, she finally came to the sensible conclusion that her son could fare no worse as a scholar than he had as a farmer, and with a gesture of resignation—shoulders hunched, eyes pinched into a network of fine lines—whose true meaning is now lost to time, she gave her reluctant permission, striking evidence, in this case, at least, of the power of passive resistance.

Isaac Newton entered Trinity College at Cambridge University in the summer of 1661. The English civil wars had recently been concluded, one Charles separated from his head, another now on the throne, the great Puritan frenzy receding into the past, like a fever that breaks and then is gone. Newton was eighteen years old, the hot bubble of his genius rising, and despite the fact that his mother's family was by now reasonably well off, he entered the university as a subsizar, which is to say, he was expected to wait expectantly on one of the senior fellows, fetching his books, cleaning his clothes and room, emptying his chamber pots, and in general occupying himself in a variety of unpleasant tasks. Newton was assigned to Humprey Babbington, which was a stroke of good fortune, for Babbington was in college for only a few months each year. Those chamber pots empty, Newton was left to tend his thoughts.

The first year or so, he spent isolated and alone, temperamentally at odds with his roommate and the other students, if only because they regarded their tenure at Cambridge as a splendid opportunity to spend long hours in the village taverns, drinking into the late hours of the night and courting the local barmaids, while Newton, fluent in Latin and little else, saw in his narrow escape from the plow an unfilled opportunity, his immense restless intelligence wandering over seas whose waters he could hear but whose currents he could not yet discern.

The record of his studies is incomplete. Cambridge University was in the mid-seventeenth century still officially consecrated to a study of Aristotle and the Aristotelian corpus, the scholarly commitment having been a part of its original charter in the fourteenth century. This was yet a magnificent cathedral of thought, one that had been elaborated with great skill by the scholastics of the twelfth and thirteenth centuries, but cracks had long appeared in the cathedral walls, and by the seventeenth century bats could be seen flitting in its belfries. Aristotle had given the western world its essential categories, its distinctive ways of organizing experience, but his system lacked two ideas that would prove necessary for the development of mathematical physics. It was qualitative and not quantitative, and it made no provision for experiment. Twentieth-century philosophers have sometimes suggested that Aristotle was temperamentally indifferent to detail. His unhappy conviction that women have fewer teeth than men is often offered as an example. Quite the contrary. Aristotle was a superb naturalist, but the facts he commanded were limited by the tools he possessed. He had access neither to a microscope nor a telescope; and the very idea of an experiment that might force a choice between competing theories lay more than two thousand years in the far future.

By the early sixteenth century, parts of Europe were tense with new ideas in philosophy, mathematics, and what was then called natural philosophy. An air of expectation and innovation prevailed as for the first time a civilization began the work of amassing a detailed and precise body of facts, the living record of reality. One hundred years earlier, the great artists of the Italian renaissance had come completely to command the laws of perspective; thereafter they were able to reproduce the real world in unheard-of pictorial detail. By the time that Leonardo Da Vinci was able to record his thoughts, the closed system of the ancients had opened into an infinite vista, Da Vinci's insatiable physical curiosity exploring a universe of possibilities that Aristotle did not know and could not have conceived—helicopters, submersible boats, prodigious war machines, cannons and catapults, the precise anatomy of a bird's wings, techniques for casting bronze. Working in secret, but working nonetheless, Dutch anatomists had revealed aspects of the human body invisible to the Greeks and a mystery to every other culture. When in the seventeenth century, Rembrandt painted *The Anatomy Lesson*, he was conveying the details of a procedure that had already passed into maturity. Whatever else those dour Dutch doctors were doing, they knew where in the human body to look, and they knew what they might expect to find. The diapason of life was changing, the European culture coming into being now marked by a single dominating attitude of curiosity.

But if the diapson of life was changing, it was changing elsewhere. Cambridge appeared even to its own scholars as a backwater, dreary in its outlook, and more or less a place where ambitious young men with few prospects in life could shuffle along a secure ecclesiastical pathway, securing ordination and thereafter a sinecure of sorts, a life of ease in which, if nothing else, a man's supper was at least decently secure. If it was necessary to study Aristotle and accept without question a world view that any thinking person could see was declining into desuetude, Cambridge had any number of young men prepared to do just that.

The curriculum that Newton was supposed to study he seems largely to have ignored, or to have examined with indifference; but in that odd fashion in which powerful minds seem capable of splitting themselves along axes that later generations are apt to find incomprehensible or bizarre, Newton spent long hours studying alchemy and longer hours still studying Biblical chronicles, subjects that lodged themselves into the budget of his permanent interests. When late in life he found himself drenched in honor, his scientific work complete and the great engine of his curiosity turning fitfully, he returned to the Biblical chronicles, drawing up immense genealogical charts, dividing precisely the flow of ancient time.

It is almost impossible to know just what he read or studied in any detail, and so the shape of his early influences can at best be partially reconstructed. There is no record, curiously enough, of an encounter with an electrifying mentor. Newton never discharged in veneration any part of the immense intellectual anxiety that was as natural to him as breathing. His official tutor, Benjamin Pulleyn, remains in history as a cipher, a man content to let Newton do what he felt like doing, and if, on those occasions that Newton actually presented himself, Pulleyn was disposed to question Newton's readings or interrogate him closely on the arguments he had developed, something in Newton's coal black eyes must have persuaded him that it might be the better part of wisdom to let this young man move quite on his own.

In mathematics and in natural philosophy, Newton was self-taught, the last great figure on the European stage to have withdrawn to his chambers and then emerged with a soul-shattering world view. Advised to study geometry, he quickly determined that the first fifty propositions in Euclid's *Elements* were obvious, and like all born mathematicians, he saw no need to demonstrate what he could at once discern. Euclidean geometry is nonetheless a commanding intellectual structure, the first, and perhaps the greatest, axiomatic system; and while Newton may have quickly satisfied himself that the propositions of Euclidean geometry were true, some sense of the structure's grandeur quite plainly seeped into his soul, for when years later he collected all of his scientific thoughts in a mighty gesture, he chose to model himself after the Master, his assumptions carefully marked, definitions carefully expressed, and conclusions carefully derived.

At some time during the cold wet Cambridge winter, or the tentative spring that followed, Newton acquired a copy of Descartes' *Géometrie*. It was a work that inflamed his imagination. No wonder. The *Géometrie* is an inflammatory work, igniting geometry and algebra by the expedient of combining them. Within Euclidean geometry, the plane lies flat and unformed, the background against which various geometric figures repose. Against the everlasting flatness of the plane, Descartes laid the cross of a coordinate system and so made possible for the first time the invigoration of space by means of number. The coordinate system that results depicts two perpendicular lines marking distances in four directions.

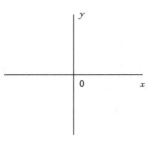

Cartesian Coordinate System

With numbers assigned to the axes of a coordinate system, Descartes realized, pairs of numbers could as well be assigned to points, as on a map. The plane then opens itself to algebraic manipulation, the movement of lines and curves corresponding to certain equations. Space for the first time comes under the domination of a symbolic system.

This is to remain within the realm of pure mathematics. Analytic geometry made possible another prospect. Algebra turns on the notion of a variable, a letter such as x that successively takes different numerical values, so that $5x$ may be 5, 10, or 50, when x designates 1, 2, or 10. There is in this capacity of a variable to take successive values some tantalizing, obscure hint of a way in which to bring change and thus motion within the powers of a symbolic system. A moving object, after all, is moving in virtue of some change with respect to some fixed place. Euclidean geometry treats places as places; but analytic geometry treats places as *numbers*. And if variables can record a change in numbers, they can as well record a change in place.

This is something that sitting in his room at night, the windows closed and the air fetid, Isaac Newton must have known.

❋   ❋   ❋

AT SOME POINT during his early years in Cambridge, Newton purchased a notebook with the aim of recording his thoughts in some ordered fashion. Such is the *Quaestionaes quaedum philosophicae*—the questions on philosophy. The notebook is divided into some forty-five sections, each dealing with an aspect of natural philosophy, the headings valuable, of course, for the clues that they provide to Newton's reading. There are entries on motion, time, space, the cosmic order; a number of entries deal with specific properties of things, their tactility or malleability, for example. There are some very effective criticisms in the notebook. Descartes had argued that light is a form of pressure, something transmitted by vortices; sight is the response of the optic nerve to a contact force. If so, one should be able to see as well at night as in the day; but one cannot. Newton's argument is clear, vivid, brief, and devastating.

Much of the notebook is of no general interest, but in some respects Newton's remarks vibrate with ideas that go beyond themselves. There is throughout a tone of urgent questioning, a kind of directed interrogation. This suggests that quite unconsciously, Newton had already absorbed one of the crucial metaphors of the scientific era, the comparison of nature to a book or text. This metaphor has guided scientific inquiry, for a long time, certainly since the beginning of the seventeenth century, and like all metaphors, it seems to have an aspect both of inevitability and timelessness. Who could doubt that nature is a book, one best read by specialists prepared to master a foreign language?

The deflationary answer is that this powerful metaphor is itself an invention, one that came to human consciousness over the course of no more than one hundred and fifty years. Nature is no more intrinsically a book than the sun is intrinsically a god, appearing and disappearing each day in the course of his appointed rounds. In the early twenty-first century, we have not entirely made our peace with the meaning of the metaphor, treating the laws of nature as answers to certain textual questions without asking too closely whether the grand image of nature as a book might not carry certain implications we are not disposed to examine. Books, after all, are written, and they carry meaning, and these are claims about the cosmos that go far beyond the laws of nature.

Newton's *Quaestionaes* are interesting in another respect. Although the notebook is divided into sections dealing with separate topics, it reveals nonetheless the imprint of a powerful unifying urge, almost as if nature were rather like a cathedral.

Just as the metaphor of nature as a book was to have a profound influence on the course of western science, so the correlative metaphor of nature as an architectural structure, one whose every aspect could be understood in terms of the relationship of part to whole, was to play a deep, if often unconscious, role in the direction later taken by the physical sciences. The primacy that physicists attach to particle physics reflects their pride that like most structures, nature is built from the bottom up, and that *at* the bottom, one is apt to find particles and particle physicists. The contrary thought that nature has *no* intrinsic organization remains on the margins of subversion, half-hidden in shadow, and not yet ready to command our assent however much it has already commanded our anxieties.

※   ※   ※

CAMBRIDGE UNIVERSITY sits on low-lying country; the fields are flat and the sky diminished. The university is now exquisitely maintained, but it retains among undergraduates and professors alike a reputation for unhealthiness, colds and various forms of grippe settling in with the winter rain and persisting for months thereafter, as red-nosed scholars scuttle along the footpaths bisecting the elegant gardens, honking noisily into their handkerchiefs and thereafter snuffling surreptitiously during lectures. His own nose raw and the winter damp seeping into his clothes, Isaac Newton spent his first university year unhappy and alone. Students who came from wealthy backgrounds and could afford their own servants must have regarded the sallow Newton as nothing more than a provincial oddity, someone who hardly spoke and who seemed forever lost in thought.

At some time in the year that followed, Newton met Nicholas Wickham, a Cambridge student offended by the raucous conviviality of his own roommate, and after a satisfying exchange in which each young man complained of his "disorderly companions," Newton and Wickham discovered that they had been drawn close by the curtain of their indignation. They took rooms together, organized their lives soberly, and if Wickham had any idea of the smoldering intellectual power commanded by his roommate, he evidently kept his suspicions to himself.

And thereafter, much of Newton's undergraduate life is lost, at least for a number of months. At some time in 1662, wracked by some spasm of guilt, Newton composed a list of sins, the composition evidently intended to do what all confessions are intended to do, and that is place a man at some distance from himself. Newton had spoken disrespectfully to his mother; he had slapped his sister; he had indulged in frivolities on the Sabbath. These peccadilloes may have troubled his conscience, but they did not blacken his heart.

At the same time, we know from anecdotal accounts that these powers of self-laceration were matched in their intensity by an almost voluptuous intellectual energy, one amounting to mania. Newton's day-to-day life was disorderly; he slept when sleep overpowered him. He read continually; he meditated without pause, often for days on end. He was often too distracted to eat, fellow students apparently regarding this as an ineradicable sign of personal perversion. He moved within a world he was unwilling to share and unable yet to master. This was now his life. It was what he had been given.

As the short brittle English day collapsed without protest into night, two tall, sober, somewhat spotted English adolescents, dressed in disheveled black, their stockings bunched about their ankles, walked somberly along the narrow gravel footpaths that led from commons to their own quarters, their conversation, one suspects, low and unforthcoming, Newton dragging his words from the deep interior place in which he kept and managed his thoughts, Wickham responding somberly in assent. Like all men whose chief bond is their dislike of others, they were bound together in a conspiracy of silence. Their friendship was to last for more than thirty years, Wickham gradually undertaking to become Newton's amanuensis and his assistant, his companionship shedding some purely human warmth into an atmosphere made chilly by his roommate's deep indwelling nature. If Newton wished for anything more than Wickham's familiar comforting homely English face as he took tea in the late afternoon, or trudged off to take his meals, it was a desire he did not express and a need upon which he did not act.

⤙ 3 ⤚

THE INFINITE

AT SOME TIME IN THE SUMMER OF 1665, officials in London discovered the bodies of two French sailors in an alleyway by Drury Lane. There was in this nothing unusual. Life in seventeenth-century London was both short and cheap; the dying died where they had dropped. It was the manner of their death that prompted concern. Both men had plainly died as the result of a swift, savage infection; and both men were riddled with buboes, ugly purple swellings that had erupted most generously in their armpits and their groins. The diagnosis of plague was unmistakable. In the fourteenth century, the plague had killed more that one quarter of Europe's population, sweeping up from Italy like a dagger stroke, emptying villages, destroying local economies, darkening and mystifying the best minds on the Continent as the disease proved resistant both to medical intervention, in most circumstances unsound, and prayer, in most cases unavailing.

The plague had returned to Europe twice more, each time

with diminished severity, but in a world still baffled by infectious disease, and so much like our own, authorities in England could think only of reposing their confidence in distance as a way of dealing with an affliction they could not otherwise control. Cambridge University closed its doors, its scholars retiring discreetly to the countryside, its lawns and gardens and stone buildings left in the care of servants who had nowhere to go and the few students who had no means to go anywhere.

Newton retired to the countryside at Woolsthorpe, four years' worth of education and a bachelor of arts degree in his possession. The sixteen months of passionate and ecstatic study that followed are by common consent regarded as comprising a year of miracles (*annus mirabilis*). As the plague made its way through the English cities, Newton, solitary and alone, decisively advanced beyond other European mathematicians, making himself first their match and then their master. In old age, his powers weakened, Newton recalled the year in which he surrendered himself to the fires that had until then smoldered in his imagination, and in a grave, magnificent, but characteristically dry description, remarked that he "was then in the prime of [his] age for invention and minded mathematics above all things."

In coming to mathematical maturity in the second half of the seventeenth century, Newton was fortunate; indeed, he was lucky. There are times when even within mathematics, nothing happens. The clock continues to tick as it has always ticked; the shadows on the sundial lengthen and then contract. Apart from what is known, nothing is new. The skin of the subject shrivels. Mathematicians amuse themselves by arranging and then rearranging the same handful of books. It rains. Someone coughs wetly in the humid night. And then all at once, things change. No one quite knows why. Mathematicians who had been snoring wake with a snort and their eyes still red with sleep, begin to riffle through their papers, noticing and noting, drawing connections, framing conjectures, scribbling their now urgent messages onto the narrow margins of the page.

The mathematicians whose chattering voices Newton heard had been wide awake for no more than one hundred and fifty years; they had mastered decimal notation and invented the logarithm. They had created analytic geometry and learned to solve cubic equations and they had begun to classify algebraic curves. The subject they created resembled an archipelago in which particular problems thrust themselves up from the ocean floor. The land bridge that binds those problems together and shows the mathematician that they are separate aspects of a common mass had not yet formed or had formed incompletely. Mathematicians amused themselves by asking questions. They reveled in their cleverness. "We are well assured," Johann Bernoulli wrote in a letter posing a famous problem,* "that there is scarcely anything more calculated to rouse noble minds to attempt work conducive to the increase in knowledge than the setting of problems at once difficult and useful. . . ."

If the great mathematicians of Europe saw islands emerging in a sunlit sea, the land bridge that connects them *was* nonetheless forming itself, binding various parts of the whole in places, its massive presence recorded by means of something like an anticipatory tingling among the more sensitive

---

*The problem of least descent, in fact. See pp. 161–162..

mathematicians, a kind of awareness that the shape and nature of the subject was changing. The seventeenth century, despite bad roads, poor food, appalling sanitation and deplorable hygiene, was the era in which the mathematical imagination transcended itself in a magnificent and deeply puzzling act of self-creation.

Some thirty years before Newton came to mathematical maturity in 1664 or 1665, Galileo remarked that the book of nature was written in the language of number. In this, he was repeating a theme familiar since at least the time of the ancient Greeks. But while philosophers and mathematicians had over the course of two thousand years discerned that there was a connection between numbers and a world of material objects, they did not possess the tools to express an insight as much mystical as mathematical. As the years of the seventeenth century consumed themselves in war, regicide, and idiotic religious controversies, the mathematicians began exuberantly to collect the techniques that they would later extend into a mighty theory of space and time, making good finally on the fateful promise of their art. The calculus was their supreme creation, the grand synthesis by which modern physical science would be made possible. And while parts of this synthesis had been at hand for more than one hundred years, it fell to Isaac Newton and Gottfried Leibniz, working quite independently, diligently to assemble the scattered clues and in an act of superb self-control, see the point where the connection between disparate concepts was to be forged.

Newton hardly seemed likely to dominate the world of mathematics. Whatever the lectures he attended at Cambridge, so far as we know, he attended them in silence. He knew Euclid's *Elements;* he had read Descartes' *Géometrie;* he certainly understood the elements of seventeenth century algebra, analysis, trigonometry, and just possibly something of the theory of probability, still an incomplete discipline, the record of vagrant hunches. He had with poised pencil read the works of William Oughtred, François Viète, and Frans van Schooten, and he had studied John Wallis's *Arithmetica infinitorum*. But although self-taught, he suffered neither from intellectual insecurity nor a tendency toward irresponsible bravado. His appreciation of seventeenth-century mathematics was from the first generous and mature. Nor was he inclined to underestimate the time and thus the effort his education had taken. In a letter written in 1695 to Nathaniel Hawes, the treasurer of Christ's Hospital, Newton, commenting on a curriculum that he had himself in part proposed, remarked that "I reckon two years too short a time for this scheme of learning."

In the first twelve months of his countryside exile, Newton ranged over the entire body of seventeenth-century mathematics. Among his papers, the Tract of 1666 offers a partial but nonetheless tantalizing record of his progress, the play of his mind. It was the idea of continuity that dominated Newton's imagination. This is an idea that arises from the experience of life itself. However much time may be broken in waking and dreaming, or work and leisure, something irresistibly suggests that these are artificial stages in what is fundamentally a primordial flow, a movement proceeding sedately but proceeding inexorably from the past into the future. If continuity is the very mark of experience, it is a mark as well of the way in which material objects behave. The sun rises and then it sets, its passage unbroken into aspects or parts, its movements proceeding without pause in a great arc;

the moon waxes and wanes, making its way through the hooded sky in a chilly haze. Projectiles leave the cannon's mouth and sail upward until they reach their apogee and then smoothly and without interruption turn themselves around and return to earth. In all of this, a sense of seamlessness predominates, almost as if the sun, the moon and the cannon's weighed charge were being carried by the same river carrying consciousness itself, their differences in speed or direction subordinated to similarities in the very nature of the process.

Newton's intelligence was abnormally sensitive to the nuances of continuity; he thought in great languid movements, his mind turning to flows and fluxions, things changing smoothly; he was able apparently to see geometrical shapes deform themselves continuously, as when a circle is flattened to become an ellipse. And he was one of the first of the great mathematicians to subject the experience of continuity to the discipline of number.

Curiously enough, the concept of continuity does not itself figure in his thought. It is the presumed background, playing even in this case the role that it plays elsewhere. Newton never succeeded in forcing his own powerful sense of continuity into consciousness. The mathematical tools that he made his own would come to make complete sense only against a background clarified two hundred years later, by mathematicians such as Cauchy, Weierstrass, Dedekind, Kronecker, and Cantor, who knew what Newton had done and could see where the edge of his self-consciousness had ended.

Newton's characteristic mental motion was a smoothly executed slide in which he would gain purchase somewhere amidst the ordinary numbers and then, with a tremendous burst of acceleration, take off for points beyond. The binomial theorem offers a striking example both of the force of his mathematical intelligence and its natural trajectory. The theorem addresses a simple problem: if two numbers are added, and their sum is then raised to some power—$(a + b)^2$, or $(a + b)^3$, say—what then the result? Mathematicians before Newton knew how to compute $(a + b)$ to arbitrary integral powers. What they lacked, those mathematicians, was a way of expressing their knowledge in a single line of concentrated code.

Enter now Isaac Newton, but twenty-four, newly retired to the countryside, never clean, never personally scrupulous. There is ink in the crotch of his fingers. Books on the table in disorderly stacks in Latin, Greek, French, and English, the heaped and careless treasure of European learning. The windows are closed, the air unmoving, the bed unmade, Newton's disheveled linen piled in a corner of the room and the remains of supper congealed on a pewter plate left standing on the top of a walnut armoire. I believe the chamber pot has not been emptied. Newton moves from his bed to his chair, entwines his long fingers, stares into space, and then when the gust of his thought has blown through his mind, bends over and grasping the ink-dipped quill in hand, begins to cover a new sheet of parchment with small crabbed numbers.

That concentrated line of code emerges:

$$(a + b)^n = a^n + na^{n-1}b + \frac{n(n-1)}{2!}a^{n-2}b^2 + \ldots + nb^{n-1} + b^{n*}$$

---

*There is only one unfamiliar symbol here, and that is 2! Read "two factorial," it signifies the product of two and one. By the same token, $5! = 5 \times 4 \times 3 \times 2 \times 1$. And ditto for all the rest.

But there is more. The binomial theorem covers all cases in which expansion proceeds by an integral exponent. The theorem makes no provision for negative or fractional exponents. It is here that the sturdy door of the familiar opens to reveal entirely a new prospect beyond, a modest enlargement of an expression's scope engendering a subtle and profound reassessment of the strategies for its expansion. Where before, a single finite formula carried its weight no matter the numbers, now the requisite formula must be expressed in *infinite* terms, the prosaic world of symbols and signs exploding outward.

Newton's enlargement of the binomial theorem made use of mathematical expressions known as *infinite series*. An infinite series is just what the term might suggest, a series of numbers that goes on forever. Consider thus $(1 + x)^n$, where $n = \frac{1}{2}$, and $x = 0.1$. The symbols embody a request for the square root of 1.1. But no strictly finite formula suffices for the computation at hand. Instead, there is this:

$$(1 + x)^n = 1 + nx + \frac{n(n-1)}{2!}x^2 + \frac{n(n-1)(n-2)}{3!}x^3 + \ldots +$$
$$\frac{n(n-1)\ldots(n-r+1)}{r!}x^r + \ldots$$

an *infinite* operation conveyed by finite means. The sums in the series are evaluated one after the other and the value of $(1 + x)^n$ given in stages: $1 + \frac{1}{2}(0.1)$, $\frac{1}{8}(0.1)^2$, and on and on. The further the expansion, the better the approximation to the square root of 1.1.

This is in itself exciting, rather like watching an unsteady series of images snap suddenly into focus and then as they move along the projector's ratchets reveal an unsuspected scene; but an infinite series does more than simply convey the mathematician from step to step. As the partial sums accumulate, they may well acquire a limit, a number toward which they are tending. If so, that number is assigned to the series as its sum, this completing the domestication of the infinite. The expansion of an infinite series offers the mathematician slices of the truth; its limit brings the mathematician the truth itself.

This Newton knew and understood, of course; he was a master of infinite series. But the concept of a limit, like so much else, Newton could not define. Nor could anyone else. And this, too, is an aspect of his own occluded self-consciousness, for the definition of a limit required mathematicians to undertake a profound and concentrated two-hundred-year meditation, the concept of continuity itself finally emerging into definition like the sun finally moving from behind dark clouds.

If in all this, I have left the discussion shrouded in myth and metaphor, this is only because it is thus that when loitering in the seventeenth century I found it.

❊   ❊   ❊

IF NEWTON WAS fortunate in studying mathematics when he did, he was fortunate again in possessing a scientific inheritance that was brilliant, penetrating, and short. Using instruments more powerful than any known to the ancient Greeks, Tycho Brahe, Johannes Kepler, and Nicolas Copernicus had studied the heavens with unheard of patience. Their research had given the community of natural philosophers what they had never possessed—precisely organized data represented in quantitative terms.

As a result, astronomical theory for the first time came to be judged by the extent to which it predicted or explained an accumulation of detailed *facts*. Theories had, of course, always been accepted or rejected by means of the imperious standard of the truth. No one in the ancient world had ever argued the sun rose in the west and set in the east, but in no other era had astronomers or natural philosophers determined the truth to such a profound quantitative extent. It is with the accumulation of such facts that a vigorous process of question and response commenced, natural philosophers coming to appreciate for what may well be the first time that by measure and experiment, nature could be coaxed into answering questions of ever increasing depth.

By the time that Newton reached his twenties, the ancient Ptolemaic system of astronomy, in which the earth stood at the center of the observable universe, the planets swinging about its center in extravagant epicycles, had been replaced by the new Copernican vision of a universe in which the earth was one of a number of celestial bodies passing around the sun.

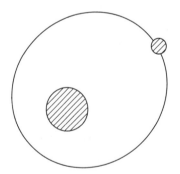

Elliptical Orbit of the Earth

Kepler had in fact done more, synthesizing his observations into three lucid and compact laws. The first simply expressed what was already evident: the planets travel in elliptical orbits around the sun. It was a conclusion that Kepler resisted, evidently because of a sense that circles were aesthetically superior to ellipses. When finally he accepted what observation revealed, he became the first man in history fully to subordinate himself to an experimental regime, and so the first of the great explorers.

Kepler's second law expressed a geometrical relationship between the sun, considered as a point in space, and the orbit of the planets, considered as a curve. An imaginary straight line connecting the sun and a planet thus connects a fixed and a moving point. That imaginary line, Kepler asserted, sweeps out equal areas of an ellipse in equal times. Equal *times*, note, and not equal distances. Kepler understood that the planets do not move at an unvarying rate of speed, slowing as they pass far from the sun, and speeding as they move toward its center.

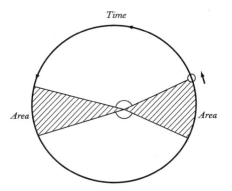

Planets Sweeping Equal Areas in Equal Times

And finally, Kepler asserted, his third law taking symbolic shape, that the time taken by any given planet to revolve about the sun, when squared, is proportional to its average distance from the sun, when cubed. Time and distance are now found to be connected by means of a very simple (although unexpected) mathematical relationship.

$$T^2 = CR^3$$

T = *Period of Revolution*
R = *Mean Distance From Sun*
C = *Contant*

Planetary Time and Planetary Distance

Kepler's laws are preeminent in the history of thought if only because they are unprecedented. They pass quite beyond observation to encompass relationships between abstract and quantitative properties of celestial bodies. Concepts lying far from common sense have been given a numerical interpretation. With the subordination of observation to theory, the human imagination has penetrated some ancient inhibiting carapace.

✾   ✾   ✾

IF KEPLER'S LAWS of planetary motion formed one part of the arch of Newton's intellectual inheritance, Galileo's laws of freely falling objects formed the other. Writing at roughly the same moment in history as Kepler, Galileo had taken as his imaginative province the motion of material bodies on the surface of the earth. His was a work both of rejection and initiation. The Aristotelian doctrine of impetus had for more than two thousand years stood as an obvious, indeed, an inescapable, account of motion. An object is thrown. It travels for some time and then loses speed as it loses steam. What better way to explain the plain facts of our experience than to suppose that some quantity of impetus or force had initially been imparted to the object? As the object moves, it loses its force or impetus and so returns to a state of rest. It is the state of rest that is natural for material bodies. And surely this is true. Rest *is* the natural state for objects on the surface of the earth; everything else involves a contrivance.

In an gesture of remarkable intellectual impudence, Galileo rejected the law of impetus; and he rejected as well the idea that material objects have a preferred state or condition. In plain fact, he dismissed the facts in plain sight, substituting for Aristotle's law of impetus his own law of inertia. An object in uniform linear motion is one moving in a straight line at a constant rate of speed. Objects in uniform linear motion, Galileo affirmed, will continue to move unless acted upon by another material object; if not in motion, objects at rest tend to remain at rest. There is no intrinsic difference between the state of motion and the state of rest. And since no impetus is lost in motion, impetus is not required to explain it either.

In comparison with the doctrine of impetus, the principle of inertia is hardly obvious. It requires an audacious act of abstraction before it may be understood, for plainly Galileo knew that arrows, javelins, horse-drawn carriages, and stately warships all go for a little while and then they stop. It is only when the same scene is glimpsed with an idealizing eye that a different story reveals itself. On the surface of the earth objects do, in fact, crawl to a stop, but *not* because they have been emptied of their impetus. They stop because something has made them stop. Absent such forces, they would proceed forever. No one before Galileo had ever seen objects proceed in motion forever; and no one after him either. Our experiences are distorted by the mask of the earth's gravity. And yet Galileo saw his way to an abstract scheme in which rest and motion lose their individuality, objects move until they are constrained, and motion proceeds on its appointed course with no impetus at all.

The same daring that allowed Galileo to see behind the general screen of appearances allowed him to see behind the particular screen that had long made a mystery of free-falling objects. Common sense would suggest that a penny and a lead weight when dropped should travel at different speeds. The lead weight is heavier. It should go faster. Not only did common sense suggest that this is so, it did so with such authority that the matter was never put to experimental test. Why question the obvious?

Nonetheless, when Galileo dropped different weights from the Leaning Tower of Pisa, he discovered that their rate of descent remained unaffected. Pennies and lead weights traveled at the same speed, reaching the ground with the same satisfying clatter. The fact that this is so is what it is—a *fact*; and the law that expresses that fact is the first of Galileo's laws of free fall.

The second follows. The speed of a falling object, and so the distance it covers as well, is determined by its acceleration and by nothing else. Acceleration is the rate of change in speed, as when with its turbines shrieking a jet plane pushes passengers backward into their seats. It is gravity that determines acceleration in free-falling objects, acting on all falling objects in a way that is independent of their nature or composition. It tugs at pennies and lead weights alike, its force changing their rate of speed by roughly thirty-two feet in every second. Gravitational acceleration is constant on the surface of the earth. It does not vary. And this no one before Galileo knew.

The discoveries made by Kepler and by Galileo were subtle and they were profound; but they were tantalizingly incomplete as well, every issue that they settled prompting another in return. Kepler had determined that the path of the planets exhibited certain geometrical regularities; he could give no reasonable explanation for their nature. Why should the planets have taken it into their heads to describe an ellipse, and not some other shape? For that matter, just why were they in motion at all? Objects on the surface of the earth remain suspended in space only briefly, and then come sailing down. What force sustained the earth in its revolution about the sun?

The air of mystery deepens when the work of Galileo and Kepler are considered jointly, for the insidiously vexing fact is that somehow Galileo and Kepler seem to be describing phenomena that in some very elusive sense are connected, if only because gravitational acceleration and planetary motion are alike in being movements of material bodies. And yet plainly, the planets are one thing, objects falling on the surface of the earth, another. The planets, after all, are not falling. Unperturbed, they simply sail through the night sky as they have always sailed.

The questions formed and reformed themselves throughout the sixteenth century; they glittered far from the place where such questions arise.

DAVID BERLINSKI

❀   ❀   ❀

ALL THIS NEWTON knew and must have known
during his year of passionate and self-absorbed study. It would
be interesting to have had access to his consciousness as it
folded and unfolded itself. His papers are by themselves unrevealing. They are rarely introspective. Yet the emergence of
a great intelligence puzzles our own imagination. Newton's
biographers all agree that the man possessed uncommon powers of concentration, holding a problem in his mind, as Lord
Keynes once observed, for weeks and months and years until
it cracked open and surrendered its secrets. This steady and
sustained ability to focus the mind is in part the common
property of mathematicians and mystics alike. So, too, certain
experiences of the infinite. After all, mystics have written
since time immemorial of an ecstatic communion with
boundlessness. Their descriptions are among the treasures of
humanity. And they are all the same. A rain washes those
suffering dry roots. The afflictions of the self vanish into a
healing void. The distinction between subject and object dis-
appears. The budded nerve endings begin to tremble. And
then the soul swoons as it is enveloped into the infinite.
Thereafter all powers of conveyance lapse, the mystic having
returned to his body able only to place a splayed hand over his
burning heart and with fingers outstretched gesture toward
the sky beyond.

   These experiences would seem alien to Newton's temperament. There is no mention in his letters of mystic communion. Nonetheless, there is this intriguing story.
Pursuing an experiment in optics as a young man, Newton
stared silently at the sun in order to study the afterimages it

provoked. It was hardly an activity calculated to improve his vision.

> In a few hours I had brought my eyes to such a pass that I could look upon no bright object with either eye but I saw the sun before me, so that I durst neither write nor read but to recover the use of my eyes shut my self up in my chamber made dark for three days together & used all means to divert my imagination from the Sun. For if I thought upon him I presently saw his picture though I was in the dark. But by keeping in the dark & imploying my mind about other things I began in three or four days to have some use of my eyes again & by fore-bearing a few days longer to look upon bright objects re-covered them pretty well, tho not so well but that for some months after the spectrum of the sun began to re-turn as often as I began to meditate upon the phaenom-enon, even tho I lay in bed at midnight with my curtains drawn.

"Like Jacob," one of Newton's biographers remarks, "Newton had confronted a god face to face and been pre-served."*

*Nous connaissons qu'il y a un infini et ignorons sa nature.***

---

*Frank E. Manuel, *Portrait of Issac Newton* (Cambridge, Mass: Harvard University Press, 1968), p. 85.

**Pascal, *Pensées*, Article III, 235.

≒ 4 ≒

THE SPECIAL INSTRUMENT

I F NATURE IS A BOOK, IT IS NOT ONE EASILY read. It is, to be sure, full of hints, like a foreign language with familiar words. Time goes from the past into the future; the seasons follow one another; the planets turn within the etched curves of the celestial sphere. These are the regularities on which life is based. But no account limited to a record of such regularities suffices by itself to reveal the secret plan by which they are coordinated. It is for this reason that philosophers have always wished to possess the special instrument that would enable them finally to read the book of nature,—the *Book*—the letters cohering into words, and then sentences and paragraphs, the meaning of the message flowing like water over dry sand.

For a very long time, philosophers believed that the instrument they required might be found just behind the familiar. Human beings bring change to the world by acting upon it. Just as a plow cuts a furrow in a field, the sun cuts a furrow in the sky. The plow is under human control, the sun moves as a god might move it. Things occur in nature as the result of the exercise of some will and that will is directed by some recognizable human emotion—fear, anger, pride, jealousy, or sexual exaltation. This way of thinking is very plausible; it may at the end of time turn out to be correct. If as in life nothing comes about without some form of intelligent agency, so perhaps in nature as well. The extraordinarily counterintuitive idea that the *Book* might not depict an emotional drama at all required thousands of years before it could emerge fully into self-consciousness; even those masterful Greeks, secure in their possession of geometry, looked over a blazing sky in which gods and goddesses crossed the heavens in their winged chariots.

It is the calculus that for the first time allowed mathematicians access to the *Book;* and in Newton's hands, it is the calculus that became his special instrument.

And ours.

❁   ❁   ❁

NEWTON'S PURELY mathematical passions burst into flame during sixteen months; by the time that he left Cambridge for the countryside, the plague in place in the university, he had already immersed himself in the free-flowing currents of European mathematics and discovered, no doubt to his satisfaction, that his powers were bounded only by his patience. It was at Woolsthorpe that those powers resolved themselves in the discovery of the calculus. Newton left no autobiographical account of his achievement; at some time during the sixteen months he spent in the country, he "*had* the method of fluxions."

41

Newton's discovery of the calculus is often represented both as his greatest and his last purely mathematical achievement. The common view is that during his years as an undergraduate and the first months of his withdrawal to Woolsthorpe, he circumnavigated the mathematical globe; thereafter his obsession with mathematics came to an end. He turned his attention elsewhere. It is very hard to believe that Newton divided his intelligence so neatly. The time in which Newton discovered the calculus was also the time, he recounted many years later, in which he began to think of "extending gravity to the orb of the moon." However much he may have passionately pursued pure mathematics, he was not in his heart purely a mathematician; during the year of miracles, it was something else that had seized his imagination, some suspected barely felt connection between that mathematical garden in which he had wandered with such careless authority and the massive collocation of astronomical and terrestrial facts that were his to touch and tame as well. Like Einstein, his spiritual heir and only equal, Newton viewed mathematics as an instrument. In thinking about the calculus, Newton was already thinking *beyond* the calculus, planets in motion and falling objects moving across the enormous corridors of his mind. Let them move and tumble, those pale planets and falling objects. Something is forming itself.

❀　❀　❀

THE CALCULUS gains its initial purchase within a Cartesian coordinate system. Like so many other mathematical objects, a coordinate system has a double interpretation and so a double life. The first appears when its axes are simply labeled by numerical variables x and y; the second, when one axis represents time, and the other space. It is thus that in one audacious intellectual act, time and space find themselves subordinated to the same system of measurement by means of the same system of numbers. In this synthesis, one sees intimations of the far future in which space and time would lose forever their distinctive identities, merging themselves into a single mathematical medium. In meditating on "problems of motion," Newton had at hand the cave or canvass on which he would paint.

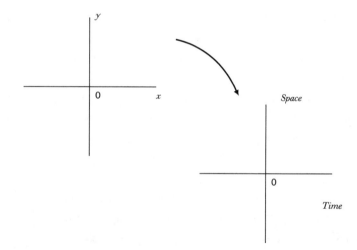

Two Labeled Coordinate Systems

A straight line is purely a geometric object. Passing through the origin of a coordinate system at an angle of forty-five degrees, it arises in the south and moves toward the north,

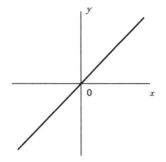

Straight Line Passing Through Origin of Coordinate System

as impalpable as the wind. It cannot be described in still simpler geometrical terms. Algebra supplies its controlling verbal artifice. A straight line moving through a coordinate system is governed by an equation whose form is $y = mx + b$. The variable x designates a number along the x-axis; the constant b, the point at which the line crosses the y-axis; and the constant m, the line's slope, or angle of inclination, commonly expressed as the ratio of differences along one axis (Δx) with respect to differences along the other (Δy). With constants fixed at $b = 0$ and $m = 1$, this line bends to the artificer's will, the equation $y = mx + b$ now specifying infinitely many points—. . . <−4,−4>, <−3,−3>, <−2−2>, <−1−1>, <0,0>, <1,1>, <2,2>, <3,3> . . . .—over the whole reach of space and time.

What holds for a straight line, seventeenth-century mathematicians knew, held as well for other curves. A parabola, dipping low to touch the origin, is controlled by an equation of the form $y = x^2$. Circles respond to algebraic command; so, too, ellipses. These are familiar examples; they only suggest the treasures beyond, for analytic geometry draws a close connection between ever more elaborate curves and ever more elaborate equations, the sinuous deformation of space undertaken by catenaries, swaying loops, and languid spirals all finding some unexpected inner voice in the arrangement of symbols, variables, and the numbers they express.

❊   ❊   ❊

TIME AND SPACE make up the world's great vault. It is within that vault that change occurs. The night sky is lit by moonlight, the great trees sway, birds shriek, and then the opalescent light vanishes as the sun appears. Something has changed because some things have changed. This is the familiar, the *continuous* world. And it is this world that Newton represented mathematically by means of a series of daring correlations. Motion is change in place. Something was there. Now it is here. The difference between here and there is simply a measure of *distance*. If in its most general terms, a Cartesian coordinate system represents a cross between space and time, the numbers marking space along one axis now represent elapsed distances from the fixed hollow of the origin. A single all-purpose question—*how far?*—resolves itself into a single all-purpose number—*this far*. A correlation now appears between distance and time, *how far* juxtaposed to *how long*.

A mathematical equation is a device binding numbers to numbers; but the very idea of a correlation admits a more general (and hence a more useful) representation in terms of a function. A function is a compact way of signifying a transformation, an exchange between what is given and what is gotten in return. Within mathematics, what is given is typically a number, and what is gotten is a number in return, and between giving and getting, a purely mathematical operation holds sway. The function $f(x) = x^2$ thus takes a number and then squares it: $f(2) = 4, f(3) = 9$, and $f(16) = 256$; the function $g(x) = x + 1$, by way of contrast, adds 1 to every number: $g(1) = 2, g(9) = 10$, and $g(256) = 257$.

This extraordinarily flexible idea allows the mathematician to describe numerical relationships by means of an antecedent vocabulary of mathematical operations.

This may seem to have little to do with geometry, but following Descartes, mathematicians learned how to displace these chattering functional exchanges so that their tense give and take came to be displayed in terms of the flowing curves that mark the boundaries of a world in space. The idea is simple. Viewed as a numerical device, a function does nothing more than send numbers to numbers; but viewed again by means of the mathematician's prismatic lenses, a function also sends numbers along one axis of a coordinate system to values along the other, leaving a living record of itself in the curve that results. This is not only lovely to look at, it is interesting as well, the mathematician once again having invigorated space by means of numbers.

The *mathematician?* No, no. A young man of twenty-four. Tall, his face angular and tense, his eyes often narrowed in intensity, but innocent in repose; and although he was hardly a model of physical grace, taking as he did no exercise beyond walking, his hands, with their long and elegant fingers, were surprisingly dexterous, capable of deftly peeling a green leaf to expose its threaded veins.

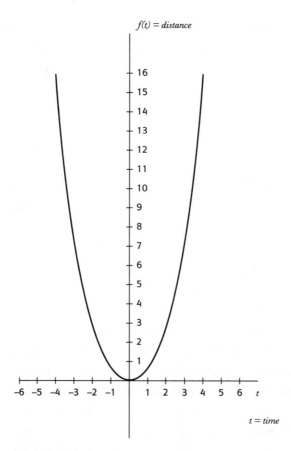

Curve Expressing the Function $f(t) = t^2$

Thus far mathematics and *only* mathematics. It is a *physical* object that is now in motion. A curve, Newton assumed, is the phosphorescent trace it leaves in space. This assumption represents a claim made against the world; it is the first clue that Newton discerned to the meaning of the *Book*. Now what moves in space moves as well in time. If a man is moving without speeding up or slowing down, the record of his change in place—or *displacement*—is expressed as a straight line.*

In turn, displacement generates a measure of speed or velocity, as when huffing along a country road, a man covers five miles in one hour.

Five miles—*distance;* one hour—*time.* Five miles *in* one hour—*speed.*

If speed is defined as a ratio, it is a ratio that has an obvious geometrical meaning. The slope of a straight line is a terse summary of change along one axis juxtaposed against change in the other. The axes now in place cross time against distance. Change in distance with respect to time is speed incarnate. The simple apparatus of a Cartesian coordinate system has made it possible to treat velocity as a mathematical and thus as a measurable quantity.

This is to remain entirely within a world of familiar concepts. But what analysis of speed commends itself if a man is in fact moving himself through space by slowing down and then speeding up or speeding up and then slowing down? A straight line in space is the record of what physicists call uniform rectilinear motion. The case at hand is otherwise.

A straight line does record change, of course, but as the drawing might indicate, it is change that never changes. Change beyond simple change requires a *curve* in order to mark its appearance against the screen of a coordinate system; it is only when a straight line bends that it acquires the capacity to depict an ever changing relationship between time and distance.

---

*Displacement is change in place; *distance*, it's numerical measure.

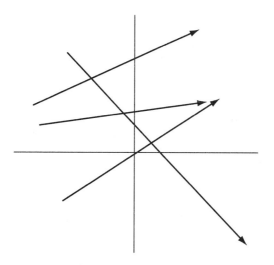

Uniform Rectilinear Motion

Now a curve in space, no less than a straight line, is a geometrical object; and no less than a straight line, a curve cannot be understood in simpler geometrical terms. It is what it is; it does what it does.

It is the art of algebra that must once again invigorate a geometrical world. If a curve in space is the geometrical trace of a moving object, a function in turn is the algebraic instrument correlating change in time with change in place. It is thus that a function's skeleton may be seen beneath a curve's face, as when the function $D(t)$—the *distance* function—correlates every moment in time with some displacement in distance. If $D(t)$ is given precise algebraic specification, so that $D(t) = t^2$, the curve rolls directly forward from the numbers, as shown in the drawing on page 47.

The request that has so far prompted these reflections

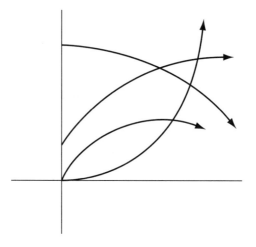

Non-uniform motion

was one made for an analysis of speed. Straight lines are useful in a world where nothing ever speeds up or slows down. Our present concerns, Newton reminds us, his dark unrevealing eyes roaming over the margins of time, are with the real world, the one in which things surge and then slow down or shuffle forward with a weary grunt and then surge dramatically. Surprisingly enough, the sparse tool of the straight line makes possible the conceptual recovery of speed in the case of curves.

Let that curve undulate where it may. The slope of a straight line joining two points on the curve—a chord or secant line—must nonetheless represent *overall* change in distance with respect to *overall* change in time. Since the secant line *is* straight, its slope is simply the ratio of distance against time. The concept of average speed now admits of mathematical expression as $\frac{\Delta D}{\Delta t}$.

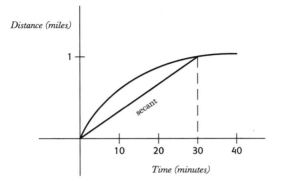

Secant Line Joining Two Points on a Curve*

Average speed is a useful concept. It is, in fact, the rough and ready concept we all use in assessing how fast we have gone when we have gone fast. But no matter how useful, the concept is also incomplete. An object in motion moves in space by prolonging distance; it moves in time by extending duration. And if it moves in space and time *continuously*, prolongation of distance and extension of time are indivisible into least parts. The coordination between space and time takes place at every point in the trajectory of a moving object.

And so it follows—a moment of command is now struck—that if speed is a physical quantity, it must as well be a physical quantity defined *at every point in time and so at every point in space.* In his *Treatise on the Method of Fluxions and Infinite Series,* Newton expressed this command as a prob-

---

*Attentive readers need not fret: the shape of the curve has changed, and this is simply for ease of illustration.

lem. "The length of space being described continually given (that is, at all times); to find the velocity of the motion at any time proposed."

To define speed at *any* time proposed, meaning at *every* time proposed, and thus to define speed as a continuous *function* of time.

❃   ❃   ❃

THE PROBLEM that Newton posed to himself embodies a request for *instantaneous* velocity and so a request for a number; but with the apparatus of distance, time, and speed, there is absolutely nothing obvious that prompts the requisite number to the imagination. The calculus gains further purchase on time and distance by coaxing that number into consciousness.

Suppose a curve given, the curve recording the continuous displacement of a moving point from the origin of a coordinate system. The correlation between time and distance is now under the control of the distance function of old, $D(t) = t^2$. At every moment, this function specifies the distance that a particle has traveled as the square of elapsed time.* Time is now flash-frozen at a single moment, t, and in the moment's flash, an odd but compelling answer of sorts to the question of instantaneous speed reveals itself, like a photograph depicting an aura. *An object in motion at any particular moment is moving as fast as it would be moving if it were moving along a straight line.*

---

*This function specifies distance but not direction. A particle leaves the origin. In the first hour, it covers two miles. It then returns to the origin, covering the same two miles. It has gone nowhere, but it has covered four miles.

The requisite straight line lies tangent to the curve, touching it just once before zooming off into space. And the answer commends itself to common sense because a straight line has a measurable slope, and the slope of a straight line is a natural measure of speed.

A line tangent to a curve provides a picture in answer to a problem; but a picture does only what a picture can do, and what is most wanted—a *definition* of instantaneous speed—is what it is least able to provide.

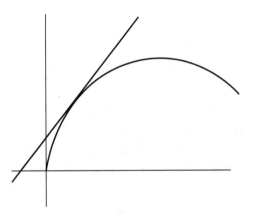

A Line Tangent to a Curve

A series of intellectual commands and half formed images now follow. Imagine an ant retiring along the back of a curve, stopping every few paces; go back now as an ant would go, and where she might stop, mark the spot. Connect that spot to the original point given. Continue as the ant marks her paces, secant lines now forming, like creases in a fan.

Let those secant lines be retracted toward the tangent

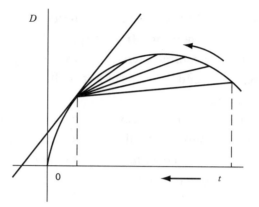

Retraction of Secant Lines Toward a Tangent Line

line, moving closer, as the ant moves closer, approaching but never completely reaching the tangent line.

Enter numbers. Each secant line has its own slope: each slope is a ratio, and so a *number*. The time between two points is now squeezed and as the temporal interval grows smaller, the slopes of those secant lines arrange themselves into a series. Taking them as a sequence, we have *the slope of the first secant line, the slope of the second secant line, and so on and on and on.* Such sequences, Newton knew, might go on forever, but *tend* nonetheless toward some fixed limit. This was an idea that he had already brilliantly exploited. And there follows a daring abjuration. If this series does tend toward a limit, let that limit be *assigned* to the tangent line as its slope.

A precise definition of instantaneous speed or velocity now drops from ordinary English. The instantaneous velocity of a moving point at a given time is just the limit of a series of average speeds.

That limit, if it exists, mathematicians designate as the derivative, $\frac{dD(t)}{dt}$, of the distance function $D(t)$.

❋   ❋   ❋

IN FACT, THE relationship between instantaneous speed and the limit by which it is defined can be circumvented. Consider a particle controlled by the now familiar function $D(t) = t^2$. After two hours it has traveled four miles; after five hours, 25 miles, and after 100 hours, 10,000 miles. What is its speed at *any* given point? As it turns out, it is $2t$. *And it is always* $2t$. If three hours have elapsed, its instantaneous velocity is six miles an hour. But ditto for the case in which six hours have elapsed. Its instantaneous velocity is then twelve miles an hour. And so to all other times.

This is an astonishing improvement both in conceptual ease and in calculation. No limits need be consulted since none are required. Instead, the mathematician proceeds from the head of a chain of definitions (distance) directly to its tail (velocity).

The shortcut that Newton discerned in this case, he discerned in other cases as well; in his hands, the calculus became a series of splendid shortcuts, the things spreading over the ground like brisk urgent little streams, shortcuts available for *all* of the elementary mathematical functions (sums, differences, powers, square roots, polynomial functions). In one of the remarkable ironies of history, it has been the shortcuts that have dominated mathematical education, so that baffled students traditionally have thought of the calculus as an efficient way of getting from nowhere to nothing. It was hardly a view to which Newton was inclined. He was among other things inordinately sensitive to the subordination of his techniques to the concepts that they served.

Among other things.

※　※　※

VELOCITY HAS been *derived* from the concepts of distance and time. By answering to Newton's demand that instantaneous velocity be defined for every moment of time, velocity has entered into discourse as a new continuous function, a correlation achieved between the tick of time and the extension of space.

This is quite remarkable. The mathematician's artifice has brought a quantitative relationship into existence. What is even more remarkable is that the artifice of definition may be self-applied, revealing the mathematical skeleton of another familiar concept. Distance divided by time is speed; but speed divided by time is *acceleration*. (It is acceleration that forces us back into our seats when a jet plane goes hurtling ever faster down a runway.) The same richly ornamented pattern prevails. There is first the fan of average acceleration, and then as the creases in the fan collapse, instantaneous acceleration, a measure of how rapidly a particle's speed is changing at a single frozen moment. The result is the continuous acceleration function, the last in this series.

And if I insist that this is remarkable, it is only because distance, speed, and acceleration are the crucial concepts that Newton required to create his system of the world.

※　※　※

THE DEFINITION of the derivative was discovered at roughly the same time by Isaac Newton and Gottfried Leibniz, his only rival on the European scene, and just possibly his only rival in all the tides of time. Each man was aware of the other's power, and each man was persuaded that in some respect it was inferior to his own. Their dispute has the unattractive quality of a conflict between large land mammals—elephants, say. If both men discovered the leading ideas of the calculus quite independently, their discoveries were not quite comparable. Expressing himself in the language of fluxions and fluents, Newton managed to conceal his insights in a notation that was miraculously maladroit. Not so Leibniz. The language of mathematics and mathematics itself are mutually sustaining. Assigning to the concepts of the calculus symbols of astonishing clarity, he created a system of notation that has proved enduring.

❀   ❀   ❀

BUT WHY NOT let the Old Boy explain his device himself. He stands there in front of Newton, a fine fleshy man in an elaborate wig.

Dear Sir, he remarks, *a sound notation is an aid to correct thought.*

Newton snorts indignantly.

The symbols $dD(t)/d(t)$ stand for the instantaneous change in some continuous quantity, $D(t)$, with respect to time. Do you see, My Dear Sir, how successfully in this case my notation coordinates distance and time so as to achieve an expression designating speed?

If Newton has seen anything, he is disposed to say nothing.

Observe, My Dear Sir, that in my notation, the relationship between the time and the function indicating distance is very clear.

Newton listens, his black eyes burning, for in his heart of heart he knows that the one gift he lacked was the gift for elegant symbols.

❧   ❧   ❧

THE DEFINITION of the derivative completes one half of a conceptual structure. The other half requires the integral, the perpetual Other in this particular drama. With the derivative in hand, the mathematician or physicist can express himself in equations, with his terse verbal formula $d(t^2)/d(t) = 2t$ abridging the English declaration that the derivative of the distance function, $D(t) = t^2$, just *is* the function $2t$. The symbolic apparatus, in fact, does more because it suggests a pattern of specification and discovery. The exchange between functions on display in this equation is known and fixed from the first. But suppose now that the identity of the distance function is not known at all, the unknown something registering its presence entirely by means of its derivative, as in the equation $dx/dt = 2t$. What is missing from this equation is the identity of x itself. There is in this expression no more mystery than might be found in $5x = 25$, which identifies some number only by the circumstance that when multiplied by five it yields twenty five. The difference is in the nature of x, which is a number in the second equation, a function in the first. Equations in which a function is identified by hints tendered by its derivative are known as differential equations; and setting up such equations is half the business of the calculus. Differential equations are the physicist's supreme, his canonical instrument, offering him a flexible way in which to describe continuous processes of the most varied sort. Solving such equations goes beyond mere specification to encompass the identification of what in a differential equation remains unknown. No easy matter. And for this, the pattern of differential specification must be reversed, so that instead of asking for the derivative of a given function, which yields $2t$ in exchange for t^2, the mathematician must ask for its *anti*-derivative instead, which yields t^2 in exchange for $2t$. The theorem that makes exchanges of this sort clear is the fundamental theorem of the calculus. It is a theorem that Leibniz knew and Newton demonstrated. And vice versa.*

❊ ❊ ❊

THERE IS IN the vast collection of memorabilia about Newton a charming picture of the manor house at Woolsthorpe.

Set on roughly an acre or two of level ground, it is two stories high, with a gently pitched roof, the stone barn attached directly to the side wall of the house. Smoke is puffing from the chimney; and a cow with what seems to be a curiously canine head stands meditatively chewing its cud a few paces away from a thin-branched, elegantly drooping tree. Seen from three hundred years, the house has acquired the naïve enchantment of the picture in which it has been captured, the childishly executed sketch conveying some aspect of Newton's

*The calculus is the subject of my own, *A Tour of the Calculus* (Vintage, 1987).

own memories of the manor in childhood. Whatever Newton's distaste for rural chores, and they were, those distastes, considerable, the manor house at Woolsthorpe was the rooted center of his life until early manhood. When he returned there from Cambridge, he returned as a scholar, and as a man of parts, and so forever beyond the plow, but in returning to Woolsthorpe, he returned home as well, and at least part of his intellectual incandescence must have been prompted by the simple human desire to demonstrate to himself if no one else that his waywardness had been its own justification. What his mother thought of her son and whether when looking at his reddened eyes she saw the tumor of an uncontrollable intellectual passion is something we shall never know. One can only imagine the dinner conversation, the Manor House lit only by candles, the light fading beyond the narrow windows, as Newton's mother, wishing for companionship more congenial than her chores, realizes, after asking Newton over and over again for the salt, that he was as disposed to make conversation as tend sheep.

The salt? Never mind.

❊   ❊   ❊

SOME FIFTY years after the year he spent in self-imposed isolation, Newton recalled the magnitude of his achievements: "In the beginning of the year 1665," he recalls,

> I found the method of approximating series and the rule for reducing any dignity [or power] of any binomial into such a series. The same year in May I found the method of tangents of Gregory and Slusis, and in November had

The Manor House*

the direct method of fluxions and the next year in January had the theory of colors and in May following I had entrance into the inverse method of fluxions.

And all this in the countryside, the animals lowing at twilight, the wind from the west carrying the smell of the sea.

*And just where is that cow? Or the smoke, for that matter?

≼ 5 ≽

NEWTON IN HIS PRIME

ISAAC NEWTON RETURNED TO CAMBRIDGE IN the fall of 1667. In leaving Woolsthorpe again, he severed his last profound association with the English countryside and a rural way of life. He would see his mother only when she lay dying.

Newton's career at Cambridge was to last twenty-seven years; in his middle years and in old age, he became a remote and powerful public figure and lived in London. Although the least sociable of men, he nonetheless ascended the academic ladder with remarkable ease, obtaining first a minor fellowship, and then a secure position on the pole of academic politics. Whatever his psychological infirmities, they evidently did not prevent him from using his superb intelligence for political purposes, and even though he often seemed to be an enemy to himself, the fact is that in all of his larger aims, whether intellectual *or* professional, he was inevitably successful.

The brief bursting fire that permitted Newton to create

the calculus had by the time he returned to Cambridge extinguished itself. His mathematical powers remained matchless; it was his interest that flagged. Many years later, he was to write that his passion for mathematics had exhausted itself "in so much that I have long grutched (i.e. begrudged) the time spent in that study unless it be perhaps at idle hours sometimes for a diversion." This was not, of course, entirely an honest judgment. Mathematics is hardly a subject that one takes up and drops. The great mathematicians are seized by their subject and so long as their powers last, the seizure lasts as well. Newton had seen mathematics for what it was—a supremely useful instrument by which he could pursue his grand scheme of physical understanding; and while he made no new mathematical discoveries after his creation of the calculus, he continued to refine his appreciation for his special instrument so that by the time he finally organized his thoughts in a single mighty effort, what had been a brilliant series of insights had become something as subtle and as flexible and as strong as silk.

His life at Cambridge was given entirely to study and meditation, and this to the exclusion of friendship, ordinary diversions, recreations of any form, and even exercise, it is pleasant to recount. Colleagues and students alike remarked on his habits, and their accounts are the same. He was consumed. When exhausted by one line of inquiry, he sought refreshment in another. His intellect was by ordinary standards indefatigable. Had he not busied himself with his hands (in various experiments) one contemporary observed, "he might well have killed himself in study."

Curiously enough, his researches were by no means entirely scientific. A good part of his time was given over to theology and to alchemy. Newton wrote more than a million words on alchemical subjects, conducting endless experiments in his own laboratory, pursuing the secret hints and whispers of ancient texts, mixing potions by hand, tasting the results, and occasionally poisoning himself as well (with tinctures of mercury, most likely), endeavoring forever to realize the scheme of coordination that accounted for chemical affinities and disaffinities.

His work led nowhere, of course; he had wasted his time. But in accounts of his alchemical experiments, Newton nonetheless left an interesting record of his characteristic scientific emotions. The alchemical literature had by long tradition been composed in an elegant but inscrutable metaphorical language (rather like the language current in modern cosmology, as it happens). The date is May of 1681. Newton's investigations had reached a climax. "I understood," he writes, "that the morning star is Venus and that she is the daughter of Saturn." The note of triumph is repeated. *I understood. I have understood. I do understand.* On May 18th, Newton remarked that "he had perfected the ideal solution." God alone knows what evil-smelling concoction he had invented. Whatever it was, it was then, and only then, that he discerned that "the eagle carries Jupiter up." This meant that "at last mercury sublimate and sophic sal ammoniac shatter the helmet and the menstrum carries everything up."*

It is tempting to dismiss both the language and the secret experiments that it reports as examples of self-deception, reassuring evidence of weakness in an intelligence notable for

---

*In describing Newton's alchemical research, I have followed Richard Westphall's account in *Never at Rest,* pp. 141–142 (Cambridge, England, Cambridge University Press, 1993).

its otherwise unassailable strength. However reassuring, this view is hardly credible. Alchemy did represent a dead end in physical thought, but in its lurid flicker Newton could certainly see the same principles that he would brilliantly exploit in mathematical physics.

At the heart of alchemical philosophy is a doctrine of chemical transmutation. If one metal *can* be transformed into another, it follows that their respective physical properties must represent the different organization of some common structure. This is hardly a frivolous thesis. It is, in fact, the very basis of the atomic theory of matter. Alchemists assumed quite reasonably that differences among metals represented superficial variations of their common chemical structure. In this they were mistaken. The common structure of complex material objects lies at the level of their elementary constituents. The laws governing the ultimate constituents of matter are not laws of chemistry at all.

Even though alchemists had failed in their larger program, they *had* by the seventeenth century acquired a body of phenomenological knowledge about chemical compounds and their various properties. The six volumes of *Theatrum Chemicum*, published by Lazarus Zetzner in 1602, although fanciful in part was also commonsensical in part, and contained a record of observation and experiment that in some ways anticipated observations and experiments that would be repeated two hundred years later.

What alchemists lacked was luck. They guessed at the principles required to organize chemical affinities, and they guessed wrong. The future lay with the new science of chemistry and ultimately with quantum physics. The great alchemical dream of the transmutation of elements would not

be realized until the twentieth century, proving, when realized, to have been uninteresting all along. The transmutation
of metals is too expensive to be worthwhile.

If Newton spent years in alchemy, he spent almost as
many years in theological research. Newton was a man of
deep religious sensibilities. Having taught himself Hebrew, he
read the Bible anew from original sources. Judging from his
papers, he seemed to have been familiar with all of the
Patristic fathers, reading them in Latin or New Testament
Greek. This represented a considerable achievement in seventeenth-century scholarship, the more so since Newton's powerful need for order was impressed on a series of disorderly
and improperly edited documents. He read with ferocious intensity, searching through the debris of ancient doctrine for
the elusive but continuous thread that would connect the initial moment of divine revelation to the various practices and
dogmas of the Church itself. He accepted nothing that he
could not verify by means of primary sources. If one thinks of
the Book of Nature as yet another text, it is easy to see that
Newton's religious and scientific inquiries were united by a
common urge and governed by common principles.

In time, Newton came to reject Christian orthodoxy in favor of Arianism. He was completely sincere and completely
confused, for in truth Newton's very independence made him
temperamentally unable to appreciate the delicate compromises that had for more than two thousand years protected the
institutions of the Christian faith.

In its doctrines, Arianism* was in fact a fourth-century

---

*Arianism is one of a number of Christian heresies; each marks a stage in the devel-
opment of the Christian church in which church fathers required enormous intellec-

heresy, and not simply an unorthodox point of view: it questioned and so attacked the mysteries of the Christian faith. God, the ecclesiastic Arius had argued, was both infinite and unbounded, his existence beyond space and time. Like all men, Christ was finite and bounded. It follows that Christ cannot share in the divinity of God. To be sure, Christ has powers greater than those possessed by men, if less than those possessed by God; but he holds those powers by divine appointment. It is thus, the Patristic fathers realized, that the floodgate to polytheism might again be opened, the Trinity revealed as an uneasy association of disparate items. What is worse, were Christ denied participation in divinity, the very doctrine of salvation would stand in jeopardy. What point the salvation of one man by another, no matter how chosen? Bishop Athanasius declared Arianism a heresy in the early years of the fourth century. His judgment was ratified by the Council of Nicea in 325. In the long light of Christian history, there is little doubt that this judgment was correct.

Newton was attracted to Arianism in part because he was compulsively driven to investigate every aspect of intellectual life as if it were his burden to commence his inquiries from their historical beginnings. He was sufficiently vain, I suspect,

tual sensitivity to judge correctly which doctrines might safely be ignored and which required unrelenting anathema. The Donatist heresy, for example, expressed the conviction that the offices of the Church could be discharged only by priests who were themselves morally unblemished. Church fathers demurred, seeing correctly that if the Church were to be a universal rock, as Saint Peter had demanded, it must then accept impurities in its vicars. The office, they declared in a gesture of remarkable sophistication, is separate from the man. In this they were again correct. The Donatist heresy was to smolder for a thousand years, erupting into frank flame during the great Gregorian reformation of the eleventh century, and again during the Protestant Reformation of the fifteenth and sixteenth centuries. Versions of Donatism are hardly unknown in contemporary political life; had Donatism succeeded, the modern corporation could never have come into existence.

so that he was unwilling to concede that his enormous efforts could result in nothing more than the ratification of his faith.

He was in addition temperamentally secretive by nature and secretive natures are quite prepared to believe that the history of human affairs is largely the history of a great many secrets. Disposed to see things that had been hidden in early Church history, he found them with ease, discovering in the Arian heresy a record of fearless speculation undermined by orthodoxy and then buried in the sands of time.

There is an additional reason that might have prompted Newton to heresy. Arianism is a radically simple doctrine. It removes Christ from the mystery of the Trinity, and so empties the Trinity of its mystery. It places God beyond the reach of time and space, revealing him only as an unapproachably remote figure, conveying himself from the Beyond to the Beyond behind the world of appearances. It strikes for metaphysical essentials and it involves no compromises.

Newton accepted Arianism as a system of belief; having assured himself that it represented the true faith, he never wavered. There is in this a profound and poignant irony. Church fathers had seen the dark disturbing future, and one thousand years after they had denounced Arianism, the circle of their anxieties closed itself in the Protestant reformation. Newton came to maturity during the great Puritan fever that swept through English life in the mid-seventeenth century and Puritanism was itself a Protestant heresy. In Arius, Newton had seen the past; in Newton, the Church fathers had anticipated their fears.

As the twenty-first century commences, we are largely unable to recapture the intensity of conviction that for all of western history has been associated with theological belief. The Newton of rational mechanics is our contemporary; the other Newton has passed beyond the grasp of sympathy. We are left with the strange retrospective image of the most powerful intelligence of the modern era sitting alone in his study at night, a candle guttering on his blackened table, the writings of the Church fathers heaped before him in manuscripts, incunabula, scattered texts, pieces of parchment, and as bats flit about the Cambridge countryside beyond the shuttered windows, busying himself in ancient controversies, his outrage rising as he realizes that buried in Greek or hidden in Latin signs of an ancient and malignant conspiracy may be found, the liberating truth his alone to discover and his alone to keep fast.

❀   ❀   ❀

NEWTON ACHIEVED first a minor fellowship at Cambridge, and when he was awarded his Master of Arts in 1668, a permanent sinecure. There is always an odd and terrible disassociation between a life as it is lived and a life as it has been recorded. For all anyone knows, Newton may have viewed alchemy and biblical studies as the burning center of his passions, treating the development of rational mechanics as an easily handled matter on the unignited circumference of his concerns. It is nonetheless the invigoration of his *physical* imagination that we read backward into his life, treating what we have taken as if it were precisely what he had given.

At some time during his early manhood and well before his genius was generally known, Isaac Newton discerned that the force of gravity might "extend to the orb of the moon." With these seven lapidary words, he initiated modern mathematical physics, his thoughts proceeding by means of a prehensile grip that relaxed its hold on one issue only when the next was at hand. By the middle of the seventeenth century, the fact that the earth was turning sedately on its axis in space was widely appreciated for what it was—a fact that prompts a question. Spinning bodies generate forces that under ordinary circumstances tend to fling off objects perilously attached to their surface.* A fly landing on a top is quickly thrown off into space; a slingshot launches a rock or pebble by means of the sling's centrifugal force. If this is so, just why do detachable objects on the surface of the earth—men and women, loose rocks, debris, animals, sand, water, even the atmosphere itself—remain pretty much where they find themselves?

It is a good question, even now, one that freshmen students of physics quite prepared to accept the earth's rotation are very often unable to answer clearly. In the early seventeenth century, the question still stood as a real, if minor, impediment in the stream of physical thought.

In fact, loose objects on the surface of the earth *do* find themselves under the influence of the earth's centrifugal force; and *all other things being equal,* they *would* fly off into space. But all other things are not equal. The force impelling objects into space is offset by the earth's gravitational force.

---

*I am speaking loosely now, as readers with freshmen physics in their past are about to remind me. Centrifugal forces are virtual forces may be eliminated in favor of a redescription of the frame of reference in which they are said to act. Thus centrifugal forces disappear entirely as separate forces if their frame of reference is itself allowed to rotate. Newton's mature treatment of lunar motion—but why give away the game so early in the book?

But the moon is different. Or so one might think. For one thing, it is far away; for another, very large, at least in comparison with objects on the earth. And yet it sails around the earth, following a somewhat irregular but entirely predictable orbit, rising and setting, waxing and waning, a body as familiar and as comforting as the sea. That the moon neither sailed off into space nor came crashing down to earth were circumstances widely regarded in the seventeenth century as meriting explanation.

As indeed, they do. Newton resolved the motion of the moon in his mind by imagining that the moon was sailing through the night sky under the influence of the *same* two forces that affect ordinary terrestrial objects, the one pulling it down, the other impelling it along a straight line tangent to the surface of the earth.

This was an astonishing hypothesis, suggesting as it did, and suggesting for the first time, that one and the same physical scheme might be powerful enough to explain the behavior of material objects both in space *and* on the surface of the earth.

Two forces. The moon would accelerate directly toward the earth like any other object, *if* its behavior were influenced by gravity and nothing else; and it would sail off into space like any other object, *if* its behavior were not influenced by the earth's gravity at all. Both forces acting together hold the moon in its orbit by more or less canceling one another.

Two *forces*. Each requires a transmutation into the world of numbers. Close to the surface of the earth, gravitational acceleration is constant, falling objects increasing their velocity by roughly thirty-two feet in each second. This is the content of Galileo's law. The centrifugal force impelling objects away from the center of the earth, Newton calculated by means of the calculus and the mechanics of circular motion.

Two numbers are now at hand, the first measuring the acceleration of ordinary objects toward the center of the earth, the second, their "tendency to recede from its centre." The ratio of these numbers, Newton determined, was roughly 350 to 1. Gravitational forces are larger than centrifugal forces. Objects on the surface of the earth stay where they are because given the force of gravity that is where they must stay.

Newton's analysis now enlarges itself to encompass the moon. The analysis of circular motion suggested to Newton, as, indeed, it suggested to everyone else in the seventeenth century, that if objects follow a circular path around a common center, the forces impelling them outward and the forces drawing them toward that common center must be balanced. Newton first determined the centrifugal forces acting on the moon using only the calculus and his own analysis of rotational forces. He then substituted Kepler's third law into the formula for circular motion. And this analysis indicated that the centrifugal forces acting on the moon were inversely proportional to the square of distance from the center of its motion.

An extraordinary inference now follows, Newton groping his way in steps toward the future. He had determined, and determined quantitatively, the forces impelling the moon to recede from the earth. He required a countervailing force.

If gravity also varied as the square of the distance between objects, the two forces controlling the orbit of the moon would be in alignment, the moon sailing unperturbed through the skies because its tendency to recede from the earth is canceled by its tendency to converge toward its center. Newton knew gravity's magnitude on the surface of the earth; he knew as well the distance between the earth and the moon. He was thus in a position quantitatively to compare centrifugal and gravitational forces, as they acted on the moon. Observing his own computations, Newton noted that "they pretty much agreed."*

These details may now be allowed decently to fade. Newton had completed two computations, answering two questions. The first revealed that objects on the surface of the earth stay put because they are forced to stay where they are; and the second, that the forces controlling the orbit of the moon are roughly in alignment. Sometime in the 1660s, Newton knew, and knew quantitatively, what forces were impelling the moon downward toward the earth; and he knew, and knew quantitatively, what forces were impelling the moon outward along a straight line. He was unable precisely to verify his theory, the detritus of detail remaining defiant; he had nonetheless completed the first profound analysis in mathematical physics.

He had come to understand the motions of the moon, forever endeavoring to recede and forever held fast.

❖   ❖   ❖

*See Chapter 10 for details. In fact, Newton's calculations did not quite work out. The moon is roughly sixty earth radii in distance from the earth. . The ratio of terrestrial and lunar gravitational acceleration should be 1 to 3,600. Newton got a ratio of 1 to 4,000. The problem lay with various incorrect units of measurements. Such things happen. Often.

NEWTON'S DISCOVERY that gravity could be "extended to the orb of the moon" was an end as well as a beginning. With the introduction of gravity as a force encompassing both the moon and objects on the surface of the earth, Newton destroyed the prospects for a genuinely mechanical philosophy of nature. That mechanical philosophy was largely the invention of René Descartes; its influence on seventeenth-century thought was general and profound. Mechanism is a philosophy constructed by means of very simple, self-evident principles. The material world is comprised of material objects. These have extension and duration. Motion proceeds in the material world by means of contact forces, one object impinging on another, as when a billiard ball strikes another billiard ball. It follows, or so Descartes believed, that there could be no such thing as empty space. If there were, material objects separated in space could not affect one another. Plainly they do. In place of empty space, Descartes conjectured that there were vortices, or lines of influence that transmitted force from one place to another. Just as a spinning rotor when placed in loose sand will send the stuff spraying, so a spinning planet will generate vortices in the material element in which it is embedded; these vortices are capable of influencing other material objects by means of the pressure they exert. Far from being an alien force, gravity is nothing more than a form of pressure. There are material objects in the universe. There are only material objects. And there is nothing else.

This is certainly a plausible scheme; it appeals profoundly to intuition. Descartes quite understood that if the analysis of change must be local, a view that the calculus enforces, then forces must act locally as well. In developing the concept of the field, modern physics has resurrected something like

Descartes' system. It is not hard to understand why. What is, after all, the alternative to the idea that forces in nature must be transmitted by means of material objects? It can be nothing other than the idea that forces act at a distance through the intervening medium of nothing whatsoever. This is the crucial concept in all magical practices, and while magic appeals to mysteries in nature, there is something in sound common sense that scruples at the idea that a material object can be influenced by a force that has not been transmitted by another material object. If forces are allowed to act at a distance, what point remains to the very notion of a *mechanical* explanation?

Nonetheless, it was precisely the concept of action at a distance that first tantalized and then captivated Newton's imagination. There is no overestimating his intellectual daring. The moon is very far from the earth. In extending gravity to its orb, Newton was filling space with mystery. Newton knew he was embracing an absurdity. He persevered nonetheless. It was a special aspect of his genius that he was able to seize the solutions that were accessible to him, while deferring, perhaps for centuries, the problems that remained.

There is in this train of thought no reflection in Newton's own words; I have spoken for him. Much later in his life, Newton came to understand the profound nature of his meditations as a young man. What he took back to Cambridge with him, then, was simply an idea. Gravity was a force whose powers might extend to the moon; it was moreover not a contact force at all, but an impalpable presence in the universe whose nature was hidden but whose form was revealed in symbols.

❁   ❁   ❁

DAVID BERLINSKI

NEWTON RETURNED to Cambridge having confirmed his genius to himself. Had Newton been entirely as myth describes him, he might have remained a gem whose luster shown only from the ocean floor. He used the singularities of his personality to wonderful effect. By 1667, Newton had collected his thoughts on infinite series in a long manuscript—*On the Analysis of Infinite Series.* It did not contain Newton's complete account of the calculus, but it did demonstrate a magnificent mastery of infinite series. Had it been known, it would have made Newton's reputation. The manuscript came into the hands of John Collins, an enthusiastic mathematical amateur, and a man of instinctive generosity. Collins took it upon himself to circulate the manuscript to men prepared to appreciate it. The precise details of these exchanges are now lost, but when he was quite certain that word of the manuscript would have had its desired effect, Newton, understanding intuitively that a whisper is often more suggestive than a shout, withdrew it from publication.

When sometime later Isaac Barrows, who had once taught Newton mathematics, determined to resign the Lucasian professorship at Cambridge in order to pursue theological studies, he understood that only one man at the university might succeed him. It was thus that in his twenty-seventh year, Newton became a professor of mathematics at Cambridge University. He was free to do as he pleased and to think as he would.

The days follow, then the weeks, and the years, the low flat English light breaking apart the night, birds in the trees, the countryside, even in Cambridge, spreading its early morning fragrance through the streets and gardens of the university. Newton rises at six, disheveled, splashes water from a bowl on his fingertips and wrists, and after pausing to look from his window at the breaking day, repairs immediately to his blackened desk where without interruption he continues to chase a thought he had chased the night before, writing in his neat clear hand, the last traces of sleep utterly gone, his mind free, focused, flexible, racing calmly. His roommate, John Wickam, enters his chambers and nods. The two men have little need to chatter. Newton rises and fetches his gown and together, somber as two shades, they descend the narrow staircase and exit their chambers into the morning light. Tea and coarse bread or some horrid English porridge for breakfast. A slow sedate meditative walk back to chambers, the sun now ascending into the eastern sky, Newton pausing by the gravel where he can see the very diagrams he had inscribed weeks before, carefully preserved by fellows already mindful that Newton's diagrams were the effusions of a genius that they could not fathom and had no wish to disturb.

⚖ 6 ⚖

THE FIELD OF RANCOR

T SOME TIME DURING THE 1670S, THE Lucasian professor of mathematics must have entered the lecture hall at Trinity College, deposited his notes on the lecture table, and with the inward and contained sigh that all teachers sigh, begun to address a crowd of slouching undergraduates. He was by every account an indifferent speaker, staring into space as he talked, and allowing long pauses to accumulate between his sentences. His announced topic was optics. He had revolutionary ideas to express and he expressed them indifferently. Alone among natural philosophers of the seventeenth century, he believed that white light was composite and not pure, and he had conducted a brilliant series of experiments with prisms demonstrating the decomposition of white light into a spectrum of colors. Thereafter he had shown that a lens would resolve those colors back into a band of white light. He went further. Cutting a small hole in wooden board, he managed to direct the separated rays of both red and blue light through a second prism. The rays remained unchanged, striking evidence that these colors were indecomposable. Physicists in Europe responded with infuriating assurance that primary colors resulted from the admixture of white light with black, their assurances infuriating precisely because they had conducted no experiments to sustain their claim; and somewhere in Belgium, a group of Jesuits had written to complain that when conducting Newton's experiments, *they* could not obtain Newton's results. Beyond his views on color, Newton had advanced a novel theory about the nature of light, arguing that it was in essence a stream of particles moving in a straight line through space. In the lecture hall, no one paid him much mind and no one seemed to understand what he was saying or about to say. The hour turned; the undergraduates decamped. Thereafter, hardly anyone bothered to attend the lectures that remained. Much later, when Newton had joined the Immortals, very few of his students recalled that they had once heard him speak.

❈   ❈   ❈

THE 1660S marked a time in which the changing currents of European scientific life washed up first at Newton's feet, and then with gathering force, began to pull him from the center of his own isolation and pursuits. Word of his intellectual powers had begun to spread. Had he been left to his own devices, Newton might have well spent *all* of his time in alchemical and biblical studies. In some sense, these subjects seemed to Newton the most natural expressions of his personality, which was secretive, and his temperament, which was suspicious. And to be fair to Newton's human weaknesses, alchemy and biblical studies were no doubt easier subjects to master than mathematical physics, if only because the world that Newton needed in order to make sense of planets, particles, and forces was very largely a world he would first have to invent.

Newton's full immersion into English intellectual life came about when in 1670 he designed and then made a small but powerful reflecting telescope. The principle of a reflecting telescope had been well known for more than one hundred years. Light enters at one end, and is then reflected from a convex mirrored surface at the other. It is the mirror that enlarges the image. The result is free of the colored distortions that affect refracting telescopes. The chief difficulty of the reflecting telescope is that in order to see that magnifying mirror, an observer must look down the telescope's shaft, thus blocking incoming rays of light.

Newton solved this problem by placing two mirrors at oblique angles to the stream of light. One reflected the image onto the other, and thence to the viewer. It was an ingenious and delicate design.

With innocent pride in his achievement, Newton presented his telescope to the Royal Society in London. Now at the time, members of the Society still seemed generally unsure of its purpose, convening to hear odd lectures in which various biological curiosities would be exhibited to some excitement but to little effect. The Royal Society was nonetheless a gathering place for powerful English scientists, and if its offices could not yet discharge the prestige it had acquired, it was still an institution that invited envy from those whom it had denied admission. Newton's little telescope caused a sensation. It was compact, elegant, beautifully made by hand, and efficient, its resolving powers comparable to those of instruments several times its size. It was entirely free of chromatic aberrations. Newton was made a member of the Royal Society at once.

If membership was a source of pride, as it must have been, it was also a source of anxiety. Having become visible, Newton became vulnerable; having acquired allies, he now encountered enemies. The Royal Society was composed of brilliant and ambitious men who had collected themselves at precisely the moment when the currents of European science were gathering force. It was a moment everyone could sense, and what everyone could sense everyone wished to seize. Newton's reputation had preceded him. Some members of the Royal Society thought that like his shadow, it was well adapted to his size; others thought it entirely too large.

Almost from the first, Newton fell into controversy with Robert Hooke. With his telescope a cause for celebration, Newton took the unusual step in 1672 of forwarding some aspects of his theory of color to members of the Royal Society. A theory and a telescope present rather different targets. There is no gainsaying an instrument that works. Hooke did, in fact, believe that he had overcome the problem of chromatic aberration in a microscope, but a microscope is not a telescope, and whatever he may have believed, he could produce neither the requisite microscope nor his promised telescope. Newton's theory of color was another matter. Hooke had long considered himself a master of optics; he reviewed Newton's work with the fine patronizing sense that he was rebuking an intellectual inferior, arguing with assurance that he had already anticipated what was novel in Newton's theory and had already discarded what was inadequate.

Newton managed at first to contain his indignation. Writing to Henry Oldenburg, the secretary of the Royal Society, he said only that "having considered Mr. Hooks observations on my discourse, I am glad that so acute an objector hath said nothing that can enervate any part of it." Within months, his fury seized control of his pen. Whatever Hooke had to say on optics was not only "insufficient, but in some respects unintelligible." Students of rhetoric may well remark that these words have set the standard in academic objurgation.

Hooke was hardly alone in finding Newton's theory of light bizarre and difficult. The French Jesuit Gaston Pardies sent Newton a letter in which he outlined a number of pertinent and intelligent criticisms. Beyond these, Pardies suggested a larger issue, this one philosophical. In advancing his views, Pardies intimated, Newton had gone quite beyond observation; what he had at best was a "very ingenious hypothesis." Hooke had made the same point using almost the same language. And Newton was moved to respond.

Hypothesis? Not a bit of it.

> ... The best and safest method of philosophizing seems to be, first to inquire diligently into the properties of things, and to establish those properties by experiences and then to proceed more slowly by hypotheses for the explanation of them. For hypotheses should be employed only in explaining the properties of things, but not assumed in determining them; unless so far as they may furnish experiments. For if the possibility of hypotheses is to be the test of truth and reality of things, I see not how certainty may be obtained in any science; since numerous hypotheses may be devised, which shall seem to overcome new difficulties.

This is an immensely suggestive passage, one dialectically juxtaposing an unexceptional view of science with its negation. In order to investigate the Book of Nature one must look first, experiment next, and then, and only then, entertain hypotheses or schemes of explanation. Well and good. These are the terms of naive empiricism that have by now passed into all the textbooks.

But as one might expect, these terms are rather unrevealing. Are there, in fact, "properties of things" that may be determined quite without hypotheses? If so, what are they? A rose is red. This may seem as if the rose were simply exhibiting one of its properties with a fine indifference to theory. But if redness is a property of the rose, the judgment that the rose is red reflects a densely reticulated intellectual structure, one entirely hypothetical. *Red*—as in colored; *red*—as in blood-like; *red*—as in easily distinguishable from blue; *red*—as in rather close to violet. These judgments reflect various acts of assimilation and contrast. Without them, the observation that the rose is red would be quite empty. The distinction between "explaining" the properties of things and "determining" them, although intuitive, is also unintelligible. I suspect that Newton quite understood this, and simply chose to conceal views that would engender controversy without inspiring confidence.

Newton's dispute with Hooke smoldered for most of the decade, Hooke's animadversions coming to stand in Newton's mind for the nattering objections of an entire circle of critics. The men exchanged letters; each complained to friends about the other, Hooke garrulous and peevish, Newton remote and vituperative. In 1675, Newton sent Hooke a conciliatory but cunning letter, one which concluded with the famous trope that "if I have seen further it has only been by standing on the shoulders of giants." Historians have assumed that the flattery was a malignant reference to Hooke's stature. But whatever Newton's intentions, the letter hardly served to mollify Hooke, who *was* short, and neither man was able thereafter to deal with the other on any terms save rancor.

History has not been charitable to Hooke. He was by all accounts a man of considerable talent. He was the first biologist to observe and describe the living cell. He was an endlessly ingenious experimentalist. He had a fine physical sense. And in addition he had the unhappy ability of seeing where the worm of discovery would turn without being able to say what it would reveal.

In one respect, Hooke clearly had come to a modern understanding of lunar motion well before Newton; indeed, it was Hooke who instructed Newton and not the reverse. In a letter written in 1679, Hooke had pressed Newton for an opinion of his own theory of orbital motion. It was in every respect radically different from Newton's early views. Newton had explained lunar motion as the result of a balance between centripetal and centrifugal forces. Drawing on his own remarkable treatise, *Attempt to Prove the Motion of the Earth*, which he had published in 1674, Hooke proposed that the moon's orbit is the result of a central attractive force—so far he went with Newton—*and* the moon's inherent tangential motion in a straight line. No centrifugal forces; in fact, no balance of forces whatsoever. Left to its own devices, the moon would proceed in a straight line tangent to the earth. What impels the moon to proceed in a straight line, Hooke could not say. Nor could anyone else. Nor can anyone else. Nonetheless, under the influence of the earth's gravity, the moon bends continuously, deflected from the course it would otherwise follow.

Whatever the specifics, the exchanges between Hooke and Newton have both a manifest and a latent content; it is their latent content that is most interesting. Hooke was disposed to accept an explanation only if, in the end, its terms lent themselves to a picture or an image. Like Descartes before him— and indeed, like all of the great mechanical philosophers of the seventeenth century—Hooke was unable to discard a view in which material objects get other material objects to get going. In some decisive sense, Newton had by 1672 freed himself from these constraints. The universe that he regarded with his obsidian eyes was abstract and it was mathematical. Forces acted at distance. Particles moved ceaselessly through the vault of space. The palpability of things had disappeared.

Hooke was quite correct in his suspicions: Newton *had* turned his face from mechanical principles. But Hooke was correct in another respect as well. Whatever the success of Newton's schemes, their ultimate price in intellectual coherence would be very great. Newton had implicitly withdrawn himself from the world of appearances, reposing his confidence in principles that were very far from any sort of mechanical interaction. They justified themselves, those principles, because they worked. Hooke demurred. A scheme of understanding must furnish a picture behind the process. If talk thus of pictures is apt to seem childish, Hooke's implicit claim may be refurbished. An explanation must ultimately yield a geometrical object, for in the end, it is only within a spatial world that the understanding finds repose.

It is not clear to me that in this regard Hooke was wrong. In choosing to cast the *Principia* in geometrical terms, Newton was implicitly responding to Hooke, even if the geometry was largely an illusion. Science has followed in Newton's footsteps, physics in particular becoming ever more remote from mechanical principles. Quantum mechanics offers a view of the world that cannot be grasped in spatial terms. If prediction and the successful control of experiments is the touchstone of success, quantum mechanics must be accepted as a great and powerful theory. If it is intelligibility that is required and that is crucial, quantum mechanics, like Brooklyn, is known only to the dead.

If the dispute with Hooke was in almost all respects a clash between men who were intellectually not equal, Newton faced a rival of entirely a different stature in the German philosopher Gottfried Leibniz.* Hooke was what Newton would later call a "smatterer in mathematics"; Leibniz was by 1675 well on his way toward becoming the largest Continental personality of his time. He was a mathematician fully of Newton's stature and, in the variety and novelty of his logical thoughts, Newton's superior. By 1675 he had completed and committed to publication a version of the calculus. Leibniz visited London in 1675 in order to demonstrate his own calculating device to members of the Royal Society. John Collins indiscreetly showed him some of Newton's private papers on mathematics. Leibniz was impressed, but not surprised. What Newton had discovered ten years before, he had discovered since. During the following two years, Leibniz wrote several times to Newton, posing questions to Newton that only one master might pose to another. Rather than reveal anything of his discoveries, Newton shrank into himself, despite the urg-

---

*I discuss Leibniz at some length in *The Advent of the Algorithm* (Harcourt, 2000).

ing of Collins that for heaven's sake he make his own discoveries public. Newton's secretiveness kindled an inevitable dispute, both men arguing with complete plausibility that the calculus was a invention of their own devising. The dispute was to last until Leibniz's death. Given Newton's deep anxieties, the fact that he was now involved with an antagonist fully his equal could hardly have improved his disposition. Or his sleep.

❊   ❊   ❊

NEWTON'S MOTHER Hannah fell ill in the fall of 1679. She was by then an old woman, worn by work and childbirth. Newton returned to the countryside in order to tend her in her final illness. After her death, he spent some six months at Woolsthorpe arranging her affairs. Thereafter he returned to Cambridge. He had seen the arc of life come to earth. In the preceding twelve years, he had been transported by a consuming sense of intellectual ecstasy and tormented by insecurity. He was now alone. His overriding urge was for withdrawal and for peace. At some time in 1678 he had already declined even to carry on his normal correspondence, "for the sake of a quiet life."

As a very young man, Newton had conceived three revolutionary ideas, the first in mathematics (the calculus), the second in natural philosophy (gravity), and the third in optics (the particle theory of light). In the long years that followed, these ideas did not completely make their way into the larger world beyond Cambridge. In 1670, Newton attempted completely to organize his discoveries about the calculus by means of *A Treatise on the Method of Series and Fluxions.* It was a work he never finished. Newton's thoughts on gravity remained his own. He occupied himself with alchemy and with biblical studies. His masterpiece, the *Principia*, lay forward in the future.

In all of this, there is the strong suggestion of a personality baffled and angry by turn, fleeing from some central obsession so strong as to be unendurable. Yet another interpretation may be given to the years between 1666 and 1684. Newton had reached certain crucial insights in his early youth; he was content to let them mature. His ambitions were far larger than his achievements. In biblical studies, he proposed to address the largest questions about the nature of the universe; and in alchemy, questions about its smallest nature. He believed that the material universe consisted of material objects and forces among them, but he never hid from himself the possibility that gravity might be only one of those forces. "Whatever reasoning holds for greater motions," he remarked in a paper he intended to append to the *Principia*, "should hold for lesser ones as well." He then added a characteristically prophetic note. "For, from the forces of gravity, of magnetism, and of electricity it is manifest that there are various kinds of natural forces, and that there may be still more kinds is not to be rashly denied." Alchemy involved a search for the others. It was a search that Newton could not complete.

Nonetheless, it was the terms of the search that gave his intellectual life its remorseless drive. Rational mechanics was in place as his acquisition by 1666. He had only to complete the arch.

⊰ 7 ⊱

A Good Question

A T A MEETING OF THE ROYAL SOCIETY sometime in 1684, the astronomer Edmund Halley, the luckless Robert Hooke, and the architect Christopher Wren took it upon themselves to ask the following question. Supposing the planets to be attracted toward the sun by a force inversely proportional to their distance, what would be the nature of their orbits? It was a profound question and one that could only have been asked at that time and that place. Asked earlier, the question would have seemed utterly obscure; asked later, utterly obvious.

Curiously enough, the terms of the question are easier to grasp than its significance. As usual, there are two hands. The first is pointing to the fact that the planets describe elliptical orbits around the sun. An ellipse is a geometrical figure, and this is thus a fact about the *shape* of their orbits. The other hand now waves. An inverse relationship is one that decreases in proportion to distance. The greater the distance, the weaker the relationship. The facts about forces are facts about their mathematical nature.

Two hands. Two items. Two different perspectives. What is lacking is a connection between them and so a bridge indicating that force and shape are aspects of the same phenomenon.

It is now that the question reveals its troubling and latent depth. A connection between *what?* One hand is pointing toward a shape, which can be seen, the other, toward a force, which cannot be seen at all.

It is a connection that occasionally may be grasped and this in the only way in which we generally grasp anything—by means of contact forces that impinge on the human body. The fist lands, the shoe drops on a tender toe, the banana skin undoes our traction. Our familiarity with contact forces is hardly an explanation of their nature, but it counts for something simply because we know our bodies from the inside out and are thus in a position to say how such forces affect us, even if we cannot say why.

But the force holding the planets in their orbits is different. It is not a contact force at all. There is nothing we can directly experience. It is simply out there, tugging away at the planets, the very words—*tugging away*—revealing their metaphoric limitations by revealing their link once again to the human body. The force of gravity acts at a distance and it acts at once and somehow it acts everywhere as well.

And thus the original question returns. What would—what *could*—count as a satisfactory connection between the force holding the planets in place and the shape of their orbit?

The answer, as almost everyone quite understood, could only be immensely abstract; it could only be a mathematical demonstration, one starting with the properties of a force and concluding with the orbit of the planets. With the original question and its troubling commitment to mathematics, a new stage in the development of natural philosophy has been reached. *Adieu* contact forces; *adieu* the human body as the mark and the measure of the physical world. *Adieu* the comprehensibility of things. A different order of explanation now holds sway and is in command.

The question that Halley, Hooke, and Wren asked, they could not answer. Hooke did boast that he had undertaken the necessary derivation, but he refused to reveal the details, and so while his boast remains plausible its promise went unfulfilled.

At some time in the late summer of 1684, Halley traveled to Cambridge to put the very question to Newton. The mathematician Abraham De Moivre has provided a very well-known account of their exchange, one based on Newton's own recollections:

> In 1684, Dr. Halley came to visit him at Cambridge, after they had been some time together, the D^r asked him what he thought the Curve would be that would be described by the Planets supposing the force of attraction toward the Sun to be reciprocal to the square of their distance from it. Sr Isaac replied immediately that it would be an Ellipsis, the Doctor struck with joy & amazement asked him how he knew it, why saith he I have calculated it, whereupon Dr Halley asked him for his calculations without any further delay, Sr Isaac looked among his papers but could not find it, but he promised him to renew it, & then send it him . . .

No doubt, De Moivre's language is somewhat florid by contemporary standards—not that I have anything against florid prose; but it reveals nonetheless a moment of great intellectual drama, Newton's words—*I have calculated it*—entering at once into the history of the race as the affirmation of a mighty mathematical method.

The rest of the story is well known. Halley begged for a copy of his calculations; Newton, of course, demurred. And for good reason. When he set himself to writing out his proof, the calculations that he remembered somehow failed to cohere. He did the work again, this time successfully. In November of the same year, he sent Halley far more than Halley had asked, a short treatise entitled *On the Motion of Bodies in an Orbit.* The treatise went beyond anything Newton had ever attempted; it went beyond anything that anyone had ever attempted. Within the space of nine pages, Newton demonstrated that if the planets are moving in an elliptical orbit, they must be under the control of an inverse square force directed toward one focus of the ellipse, and he demonstrated that under the influence of an inverse square force, the orbit of the planets must describe a conic section. In one majestic gesture Newton had settled both Halley's original question and its obvious converse, and had managed to generalize his answer so that it encompassed not only the elliptical orbits of the planets, but the orbits of any object moving slowly or swiftly through a gravitation field.

Halley quite understood that what he held in his hands in November of 1684 was an intellectual revolution in prospect. It is to his credit that he did. With considerable flattery and an adroit sense that Newton's genius was matched only by his vanity, he proposed that Newton submit his work on orbital dynamics to the Royal Society. Halley's timing was fortuitous. After years in which he had kept the greater part of his thoughts to himself, Newton, as he approached the middle of his journey, was plainly determined to express himself completely and express himself in print.

There followed a strange and moving concourse between Newton and Halley, one almost unique in the history of science. Halley was a capable and highly intelligent astronomer; he had a double gift that allowed him to recognize genius without resentment and to encourage its efforts without servility. He was determined to bring Newton's thoughts to their natural end. He had no idea of the magnitude of the project on which he was embarked.

What Newton in fact delivered to the Royal Society two years later was the three volumes of the *Principia*, a work that stands to his earlier treatise as an oak to its acorn. The two years Newton spent writing his masterpiece were by all accounts years in which he was intellectually consumed. The raw massive energies that he had discharged in alchemy or Biblical studies or inconclusive mathematical tracts now came together and fused. He hardly ate; he slept irregularly; his life was lived in a circle that passed from his study to the Cambridge dining rooms and back again by means of a garden walkway. He rose early and retired late. His friend and associate, Humphrey Newton, who had replaced John Wickham in his quarters, looked uneasily on the actions of a man who had allowed everything in his life to be subordinated to an uncontrollable intellectual urge. For two years, Newton's body functioned only to sustain his mind. Folio sheets were filled with black ink as he progressed from one thought to another. Derivations were initiated and completed; proofs were discovered or remembered. Magnificent concepts were introduced and elaborated. Newton's system grew with rich dramatic power. Each bursting idea led to another and each added to an architectural structure that, like certain Gothic cathedrals, was delicate and massive at once.

In the history of science, there is no record of anything like the two years Newton spent in composing the *Principia.* Both Newton and Einstein had had miraculous years of insight and discovery in their youth; and like Newton, Einstein had spent years searching for the arch that would complete his theory of special relativity. But in comparison to Newton's ecstatic gallop, Einstein moved at a discrete trot, taking his ease from time to time, talking with friends and associates, giving lectures, even teaching. Perhaps the only figure to have experienced anything comparable was Michelangelo, suspended on his back beneath the Sistine Chapel, strong, silent, and alone.

Newton delivered the manuscript of his masterwork to the Royal Society in 1687. Halley read it with awe, Hooke with anger. The work was grand enough to affront his vanity without stunning him into silence; he remained persuaded that Newton had given him insufficient credit. That Newton had given him any credit whatsoever was an act of unaccustomed generosity. Given Hooke's reflexive charge of plagiarism, Newton responded with methodical fury, going over the manuscript of the *Principia* and striking Hooke's name at every turn. Halley was aghast; his protestations did no good. In the end, Newton succeeded in erasing Hooke's name from his masterpiece by displacing it into the stream of history, Hooke better known now for his disgruntlement than he might have been had Newton said what was in plain fact the truth—not every idea in the *Principia* was his own.

THERE REMAINS the obvious. The work that Newton undertook in the years between 1684 and 1686 presents
the biographer and psychologist with an ineradicable mystery,
one that goes beyond the sheer and irrefutable fact of genius,
a fact that we may recognize but that we cannot explain. The
Principia is without question our greatest work of pure
thought; it is now one of humanity's collective treasures. It
brings to completion the great scientific revolution initiated
by Kepler and Galileo; it contains a matchless combination of
mathematical argument and profound physical intuition. And
it constitutes a crucial demonstration of the power of certain
mathematical methods in natural philosophy. Before Newton,
no one quite believed that those methods could provide a
comprehensive system of the world; after Newton, no one
doubted it.

⊰ 8 ⊱

A STUDY IN STARKNESS

NEWTON'S *PRINCIPIA* IS DIFFICULT IN
two respects. It is written in Latin and it
is expressed in a challenging mathemati-
cal language. Translated into English and expressed in modern mathematical notation, Newton's ideas become much
more comprehensible. Even so, rational mechanics and celestial dynamics will never be counted among the easy intellectual pursuits of mankind. Their concerns lie very far from
ordinary experience and their methods and techniques are
like those involved in playing the violin inasmuch as they
must be learned by study and perfected by practice. For all its
difficulties, the *Principia* is nonetheless a work of great intellectual serenity.

Newton's system is both architecturally simple and severe.
The universe revealed by the *Principia* contains particles and
it contains forces and it contains the mathematical structures
by which they are coordinated. It contains nothing else. This
declaration must not be misunderstood. The universe is surely

everything that there is, and while the universe may contain particles, forces, and mathematical structures, it contains a great deal besides. The constituents or elements of grand opera, the time before the adoption of the Gregorian calendar, the batch processing algorithm used by the Chase Manhattan Bank, the best-loved book of modern times, and the United States are not particles and the relationships between such items cannot be resolved into mechanical forces by mathematical means. Newtonian mechanics proceeds by means of ruthless simplification and abstraction; and even within the world of particles and forces that it specifies, it very often proceeds by means of convenient but counterfactual assumptions.

If the world that Newton meant to describe and then analyze is severe, so, too, the structure of the *Principia* itself. It is composed as an axiomatic treatise, very much in the style of Euclidean geometry. Certain assumptions are made from the beginning. These are the axioms of the system. They are expressed in ordinary language, but they contain very specific physical terms, such as force, mass, and acceleration. These terms are not defined, but they are explained, Newton's explanations serving to refine and purify concepts that are often as old as time itself.

Newton's conclusions about the universe are intended to follow from his initial assumptions, just as Euclid's conclusions follow from his five famous axioms and their correlative definitions. This is the grand logical scheme. In fact, almost nothing follows in the *Principia* from Newton's physical assumptions alone. A rich body of mathematics must be interposed between Newton's premises and his conclusions before his conclusions follow logically from his premises. Newton's mathematical apparatus is as much a part of Newton's assumptions as his physical laws. That apparatus is often over-looked, especially when large claims are made about Newton's universe or even the universe of modern science; but it is there nonetheless and while there is undoubtedly a distinction between the mathematical and the physical assumptions that Newton makes, it is not a distinction easily drawn.

Newton begins the *Principia* with three specific laws of motion, two general principles of time and space, and a number of careful explanations:

The law of inertia.
Every body continues in its state of rest, or of uniform motion in a right line, unless it is compelled to change that state by forces impressed upon it.

This is a counterintuitive declaration, one unsupported by experience. On the surface of the earth, bodies at rest tend to remain at rest, and bodies in motion tend to return to a state of rest. This, Newton understood, is an interesting but idiosyncratic fact, a local oddity that has its source in friction and in gravity. In the universe as a whole, there are two *natural* states in which a body or particle might find itself. It might be at rest *or* in uniform motion along a straight line. By "uniform motion," Newton meant motion at the same unvarying speed. By a straight line, he meant a straight line. A state of rest might well be considered a limiting case of uniform rectilinear motion, one in which an object is proceeding in a straight line at no velocity whatsoever. This liberating verbal maneuver reveals a universe in which particles are in motion along straight lines, going forever from nowhere to nowhere, blank and inscrutable as the sun.

Uniform rectilinear motion is natural in both an absolute and a relative sense. A universe without forces would be one crisscrossed by particles moving in straight lines. There would be those particles and there would be the lines that they follow. This is absoluteness with a metaphysical vengeance. Only a universe in which there are *no* particles could be simpler. And such a universe amounts to nothing at all, pure nothingness having a radical if minatory simplicity denied to every other state of affairs.

By the same token, uniform rectilinear motion is natural in the relative sense that it has no explanation within Newton's system. A body in motion just keeps on going; Newton's system can explain where it is going, but it cannot explain why it is going there, nor can it explain why it is going there in a *straight* line rather than a circle, a spiral, an ellipse or any other shape. The ancient idea that there are certain preferred forms in nature may well strike us as the expression of an innocent prejudice, but some form of the prejudice would appear to be ineradicable. While Newton could and did dismiss the idea that the planets must trace a circular orbit simply because circles are inexpressibly lovely, he could not countenance a universe in which *no* geometrical figures are preferred, seeing in the stream of a straight line the spare scaffolding of the world as it appears before it has been corrupted.

If it is only uniform rectilinear motion that is natural, it follows that deviations from uniform rectilinear motion must represent an artifice of nature. A universe in which particles *change* is embroidered, and so less simple than it might otherwise be. This is again a metaphysical remark, a claim about the nature of things. Change requires time in which to change, and a universe in which time flows as things change has removed itself from the gaze of the Absolute.

❋   ❋   ❋

CHANGE WITHIN Newton's universe is, of course, change in motion. What else is there? And change in motion *does* fall within the explanatory circumference of Newton's system.

The *Principia* explains change in motion by a threefold reduction and a single law. In a universe in which particles are either at rest or traveling in straight lines, there are only three possibilities for change, rather a modest budget. A particle at rest can start to move; a particle in uniform rectilinear motion can slow down or speed up, or it can, that particle, change its direction by deviating from a straight line. Within the compass of the *Principia* these apparently disparate changes are resolved into a single Platonic Form: *all change is change in speed.* To go from doing nothing to doing something is to change speed; but so is speeding up or slowing down; and so, too, deviating from a straight line. An object that turns in a circle where before it moved straight ahead has acquired speed in a new direction.

Change in motion is the great imponderable in Newton's universe; and it comes about only when forces make their entrance. This is the province of Newton's second law:

The law of acceleration.
Change in motion is proportional to the motive force impressed, and is made in the direction of the right line in which that force is impressed.

In ordinary English, the law says that force is the product of mass by acceleration. This law Newton expressed by means of four famous symbols and two mathematical relationships (identity and multiplication):

$$F = MA.$$

Acceleration is purely a mathematical concept; mass is a matter of physics. In the most obvious sense, a body's mass is identified with its inherent power to resist change. It is what a force must overcome in order to bring about a change in velocity. In his third definition, Newton affirms that, "The *vis insitia*, or innate force of matter, is a power of resisting, by which every body, as much as in it lies, continues in its present state, whether it be of rest, or of moving forward uniformly in a straight line." This may not seem to shed much light on an unexplained concept. Innate *force?* The *power* of resisting? Of what use are these terms in an explanation of force itself? In his first definition, Newton does say that mass is a measure "arising from [an object's] density and bulk conjointly." This only leads further into the verbal badlands. If we knew the meaning of density and bulk, the cause at hand might be advanced, but the infection goes laterally, from mass to density to bulk.

The modern point of view is simpler. The mass of an object or particle, physicists now measure by comparison with a standard object, whose mass is fixed at 1 by definition. That standard object is a platinum-iridium alloy fused into a cylinder, the cylinder reposing itself at the International Bureau of Weights and Measures at Sèvres, France. This tells us how mass might be measured, but it does not quite succeed in saying what mass *is*.

The definition of mass is in any case a sideshow; it is the law itself that carries all of its own majestic weight. And it is Newton's second law that is his great contribution to natural science. It is a declaration as mysterious as it is simple. Force is the product of mass and acceleration. So much for simplicity. The mystery now follows. Is force being *defined* in terms of mass and acceleration? That would suggest that the universe contains only particles of varying mass moving at varying speeds, force disappearing as a constituent of things by means of a definitional demotion. In that case, it would make no sense to appeal to the *force* of gravity, say, in order to explain the descent toward the earth of a freely falling body. There is the earth; and there is that freely falling body. If force has been defined out of existence, nothing else remains. Material objects tend to accelerate toward one another.

Why do they do this?

They just do.

But Newton's universe contains particles *and* it contains forces. The relationship between force, mass, and acceleration is an identity, but Newton's second law is not for that reason merely a verbal maneuver. The law represents a discovery about the world, no different in kind than the discovery that water is identical to hydrogen and oxygen. That definitional demotion has now been cancelled or at least placed on hold.

If forces are genuinely a part of Newton's universe, the second law nonetheless endows them with a certain strangeness. A force acts by impelling material objects to accelerate. We can observe and measure both mass and acceleration, but we cannot directly observe or measure the force that controls them. That remains a real but inaccessible feature of the world. This is something Newton quite understood. Contact forces are easy to grasp. One object hits another. *Bang!* There it is. Newtonian forces cannot be grasped at all. They act invisibly; they act at a distance; and they act at once. It is only their trace in matter that can be detected. They have withdrawn themselves from the world of appearances, acting in the world of particles as what Newton called their "secret springs." This may well seem to represent a disturbing intellectual retreat, one moving backward to the dark night of medieval powers and potentialities. And so it is.

❀   ❀   ❀

NEWTON'S THIRD LAW has by now entered the popular imagination as the affirmation of a universal as well as a physical truth:

Law of action and reaction.
To every action, there is an equal and opposite reaction.

Obvious examples of the law are drawn from daily life. Two automobiles collide, the hood of the first crumpling the rear fender of the second. The first automobile decelerates abruptly. *Action.* The second automobile accelerates abruptly. *Reaction.* It may not be obvious that a law of nature is required to record these facts.

The significance of the third law lies elsewhere. The fact that every action engenders an equal but opposite reaction suggests a balancing principle at work in the world. So much is clear from the law's very language. Still, it is not easy to say just what is being balanced or what balancing itself means in this context. The image of a scale on which actions and reac-

tions have been placed is not entirely helpful. And yet the law
opens up a corridor to the very deepest parts of mathematical
physics, something that Newton, with his remarkable ability
to strike for the fundamentals, surely knew, or at least suspected. The metaphor of balance may safely be discarded.
Something more powerful is needed and that is the idea that
in mechanical exchanges, some quantity is conserved. Note
the twin pistons firing in this declartion: some *quantity*, something *conserved*. Moving ten people from one room into another conserves their number; rotating a coin in space
preserves its weight. The third law requires something more
abstract than either number or weight, and that is the concept
of momentum. This is a new concept, but one that may be expressed in terms of concepts already available in Newton's system. Momentum is the product of a particle's mass and its
velocity and when fully fledged it appears as a vector. To say
that momentum is conserved is just to say that when two particles interact, in equal and opposite reactions, their net momentum is zero. There is in this something almost childishly
satisfying, a mechanical resolution that satisfies a very primi-
tive artistic sense.

Conservation principles play a deep role throughout
mathematical physics, if only because the conservation of
physical quantities reflects a certain symmetry in the very
organization of the universe. And this has, of course, been
antecedently reflected in the image of a cosmic scale, a balance achieved in action. Beyond anything else, it has been
the concept of symmetry that has guided the physicist's
imagination, persuading him to pursue its trace in matter
from its largest aspects, which include the symmetries of
space and time, to its smallest, where sub-atomic symmetries

control the interaction of various strange, short-lived particles flickering into life and then recklessly extinguishing
themselves. But if the search for symmetry has tantalized
the physicist's imagination, it has also troubled his intellect,
for the fact of the matter is that we live in a magnificently
asymmetrical world, with even the human body featuring
two kidneys but only one oblate liver. Physicists account for
the world in which we find ourselves by appealing to a perfectly symmetrical universe whose initial symmetries have
been broken and now lie scattered. But if symmetries can be
broken, then something more fundamental still lies beyond
their dominion, the wheel of explanation now turning suddenly and with little traction.

❊   ❊   ❊

THE NEWTONIAN universe exists within space and
time—what doesn't, of course?—but it exists within space
and time that are absolute. Beyond anything that happens,
there is the measured flow of time; and beyond any place that
it happens, there is the unchanging vault of space. Newton's
own words express these ideas with grave force:

The law of absolute time.
*Absolute, true, and mathematical time, of itself, and from its
own nature, flows equably without relation to anything
external.*

The law of absolute space.
*Absolute space, in its own nature, without relation to
anything external, remains always similar and immovable.*

It goes without saying, of course, that if space and time are defined "without relation to anything external" then they must lie beyond experimental verification. It was precisely this circumstance that led both Mach and Einstein to question their existence. These questions lie in the future and the future has not completely answered them. From the perspective of Newtonian mechanics, absolute space and absolute time function to impress a single, universal coordinate system on the universe, one accessible from any place in the universe, and accessible as well at any time.

❄   ❄   ❄

NEWTON'S FIVE laws of motion express the *Principia's* purely physical commitments. The *Principia* is a mathematical as well as a physical treatise and the mathematics it employs expresses at least a part of its great vision. In order to coordinate distance, velocity, and acceleration, Newton assumed that both space and time are measurable. This is the fundamental metaphysical assumption not only of the *Principia*, but of modern mathematical physics. It is the assumption that endows space and time with all of its structure. The real numbers can be added, multiplied, divided, and subtracted. They go backwards and forwards from zero; they are infinite in both directions, and they have an infinitely granular structure, with real numbers appearing between real numbers. Within the *Principia*, these properties are true of space and time as well. They are true by *assumption*.

The assumption is its own reward. Without it, the *Principia* would gain no purchase on the physical world. Newton thought of space and time in terms of the real numbers because the instruments at his command—sundials, clocks, waterwheels, telescopes, magnifying glasses—all seemed to suggest a world in which neither space nor time displayed discontinuities. One minute slips into the next; one step slides into another. The inference from facts such as these to the conclusion that space and time are measurable seemed inescapable.

A complete and consistent description of rational mechanics would place mathematical and physical assumptions on the same footing, developing them both with the same attention to detail and explicitness. That work lies beyond the scope of this book if for no better reason than that it lay beyond the scope of the *Principia.*

⊰ 9 ⊱

A LOAN FROM
THE FUTURE

MONG OTHER THINGS, NEWTON WAS A
painter, the movements of his mind proceeding in pictorial steps; to retrace his
thoughts is to see the canvass on which he worked gradually
filling, until it passes from a series of deft sketches to a powerful and richly detailed representation of the physical universe.

Newton filled the world's canvass by means of his own pe-
culiar paints and brushes, creating and then using fearfully
complicated geometrical devices, the lens of his genius fo-
cussing his mind on a thousand different delicate diagrams.
They are, those diagrams, very difficult to interpret. There is
a famous video clip of Richard Feynman endeavoring to convey the physical meaning of one of Newton's more rebarba-
tive geometrical demonstrations to an audience of freshmen.
Dressed in his habitually casual style—open white shirt,

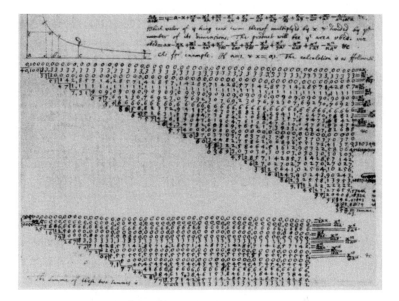

Newton's Calculations In His Own Hand

chino slacks—Feynman strides back and forth across the lecture podium, his hands flapping as he confidently barks out his reconstruction of Newton's infinitely elaborate thoughts. Then he turns to the blackboard. Diagrams go up and down. Lines begin to cross one another, the labeling ornate, but however many lines and curves and squiggles take their place on the blackboard, the argument's resolute conclusion remains one step beyond the resolving power of Feynman's own analysis. In the end, with a sheepish and endearing shrug, Feynman admits that he simply cannot make sense of what Newton has written.

It no longer matters. Newton not only created a discipline: he initiated an institution as well. If his successors lacked his surpassing originality, they did what disciples everywhere must do: they invented techniques that they could themselves handle, they demystified the master's text, and by simplifying his argument they streamlined his analysis. All of this took time. More than one hundred and fifty years were to pass before mathematical physicists and mathematicians discovered the simple, elegant tools that were to make rational mechanics a safe, sane system of thought.

❆　❆　❆

THE CALCULUS remains the crucial—the *indispensable*—mathematical tool; but it has been the enrichment of the calculus by means of vectors that has transformed the tracery of Newton's geometrical diagrams into mathematical techniques that may be used by anyone with a willingness to do a little manual labor.* Nothing really new is needed. Vector calculus is still the calculus. Limits remain as limits, derivatives as derivatives, and sequences converge to their appointed end. Vectors enlarge but to not deform the margins of understanding.

Let the universe be two-dimensional, a coordinate system imposed on its illimitable vastness, with the earth itself occupying the origin. There are thus two axes and four quadrants in space, with every point labeled by a pair of numbers. Seeing this familiar plane for the second time, the eye must now see things anew, *pairs* of numbers suddenly acquiring an independent identity of their own. To emphasize their role as objects, the mathematician places pairs of numbers in brack-

---

*Since the 1940s, even more sophisticated tools, such as symplectic geometry and differential forms, are available for the analysis of celestial dynamics and mechanics.

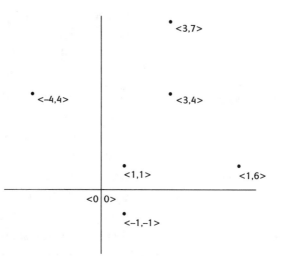

Plane With Labeled Vectors

ets. Thus $<3, 5>, <-4, 12>, <133, -40>$. And thus further to a crisp definition. *Vectors are numbers in pairs.* The parts of the pairs are their components.

Nineteenth-century mathematicians discovered that numbers in pairs could be treated *as if* they were ordinary mathematical objects and so brought vectors under the control of an ancient apparatus of manipulation. There are vectors that duplicate at a distance both the number 0 and the number 1. Thus $<0, 0>$ plays the role of 0 in the vector calculus, and the vectors $<1, 0>$ and $<0, 1>$ the role of units, or fundamental standards of measurement. Unit vectors are designated by the boldfaced letters $\mathbf{i} = <1, 0>$ and $\mathbf{j} = <0, 1>$. (Throughout vector analysis, in fact, boldface rules typographically.)

Vectors may thus be added. The sum of two vectors is the

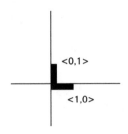

<0,1>

<1,0>

Unit Vectors

respective sums of their components. The vector sum of **u** = <5, 2> and **v** = <8, 1> is **w** = <13, 3>. Vector addition yields yet another vector. If addition makes sense, so, too, subtraction.

Vectors may as well be multiplied by ordinary numbers and by means of the same division of attention. The product of 3 and **u** = <5, 2> is simply **w** = <15, 6>. If multiplication makes sense, so, too, division. The vector **u** = <5, 2> when divided by 2 is simply the same vector when multiplied by ½. This makes for a standard description of any given vector in terms of the vector's components and its unit vectors. Thus **u** = <5, 2> is simply 5**i** + 2**j**, the original vector promptly reappearing when vector multiplication and addition have done their work. Multiplication first: 5**i** = 5<1, 0>, which is <5, 0>, and 2**j** = 2<0, 1>, which is <0, 2>. Addition next: <5, 0> + <0, 2> equals the original vector <5, 2>, as promised.

There is far more to vector analysis than this terse summary might indicate; my aim is to suggest the structure by specifying its skeleton. One aspect of that skeleton should nonetheless be clear. When vectors have been endowed with an origin and a unit, they comprise a coordinate system and so provide the physicist with a way of maneuvering throughout space. The system is entirely local. It is therefore entirely portable. One portion of space may be described by means of one system of vectors; another portion, by means of another system. The physicist thus has access to multiple coordinate systems or reference frames. These may be compared one to the other, and then by means of vector manipulation *transformed*, one into the other. Where before, one coordinate system held sway, now there are multiple copies of that original system proliferating throughout space, providing the physicist with a matchless analytical tool.

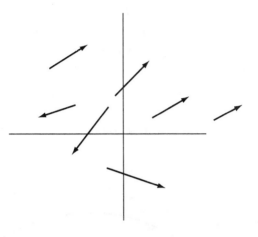

Vectors as Directed Line Segments or Arrows

In addition to their numerical identity, vectors have a second incarnation as directed line segments in the plane, or, more simply, as arrows. Although secondary, this interpretation is nonetheless invaluable. It demonstrates as only a geometrical object can the fact that vectors endow space with a linear structure. Within that stark structure, virtually every curve, and so *every form of motion along a curve*, receives an analysis, the straight lines conveying a wealth of information far beyond their deceptively simple appearance.

❊ ❊ ❊

A SPACE SHIP takes off; its trajectory is a ballooning arc through space. After a time it passes the moon and so a certain pair of points. Going further along the same trajectory, it passes Mars, and so another pair of points.

The arrows connecting the earth and the moon and the earth and Mars are the *position* vectors of the ship. They go *from* the earth *to* the moon, and they go *from* the earth *to* Mars, and unlike the ship itself they go where they are going in a straight line.

Now if arrows brings a vector to life, they bring to life something else as well. An arrow is the trace in space of distance and distance is a quantity, something measurable, a fea-

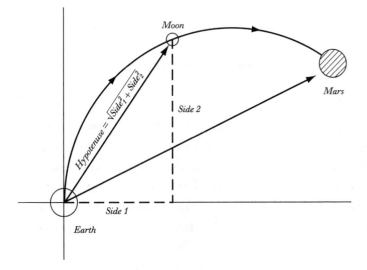

Path of Spaceship in Space

ture of the world to which an ordinary number may be assigned. This is something that a vector expresses. Given the space ship's departure from the earth and its arrival at the moon, the distance it has traveled *from* the earth *to* the moon *along* a straight line may be recovered by means of an obvious right triangle, the Pythagorean theorem, and a few algebraic adjustments. The result is a number, one expressed by a formula

$$\sqrt{Side_1^2 + Side_2^2}.$$

In a universe in which distance is crucial, distance has just made a welcome appearance.

The recovery of distance is too happy an accident to be isolated and so it is not an accident at all. Vectors are intrinsically capable of conveying *magnitude* as well as direction. This makes them supremely useful instruments. Almost all of the forces within mathematical physics go somewhere—*direction;* but they go somewhere or they do something to a certain numerical degree—*magnitude.* In vectors, mathematicians and physicists discovered one blood-filled heart with two quite separate chambers.

The ideas of time, distance, velocity, and acceleration are staples of old. They carry over to the case at hand, none the worse for wear, enriched, in fact, by their immersion into a larger mathematical system. In the calculus of a single variable, time is directly crossed against space and so makes a direct appearance on the face of a coordinate system. The coordinate system depicting the earth, the moon, Mars, and the ship speeding silently throughout the night is entirely spatial. This makes possible the recovery of direction as well as distance. Newton's law of absolute time nonetheless establishes that throughout the universe a single clock is beating. Appeals to time are directed toward that clock. So long as that clock continues to tick, it makes sense to see the space ship's trajectory in terms of the curve that it completes *and* in terms of the time it takes to complete that curve.

Vectors now recover their link to time as continuous functions. These are conveyances that take a moment and return a pair of numbers. The universal clock beats and on reaching the time t, directs the position vector *function* $u(t)$ to return with the points $<x_1, y_1>$.

With functions back in command, much else falls obediently into place. Having passed the moon, the space ship reaches Mars some time later. It has, as physicists say, displaced itself, moving from one point in space and reaching another. Displacement is simply a straight-line record of where the ship has been—the moon—and where it winds up— Mars. There is in this no more mystery than an ordinary map might afford. The displacement *from* u (the moon) *to* w (Mars) represents a change in spatial position; and it is represented by a new vector z, this one measuring the difference between u and w. Quite plainly $u + z = w$, and so $w - z = u$. As with all vectors, z has magnitude as well as direction, an intrinsic length, and so an intrinsic—an *ineliminable*—connection to the world of ordinary numbers.

Vector position and vector displacement correspond to position and change in position in the calculus. Velocity is another moth of old, emerging now as a vector butterfly. The average vector velocity of a particle is just the ratio of displacement over time, in this case $\Delta z/\Delta t$. Involving nothing more than the multiplication of a vector—z—by a number—

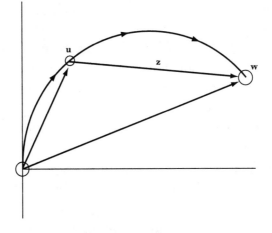

Vector Displacement

$1/\Delta t$—this is plainly another vector in turn, vectors now demonstrating a gratifying ability to pupate in space.

If average velocity makes sense for vectors, instantaneous velocity makes sense as well, as average velocities are taken to their limit. Is it odd to think of velocity in such terms? Perhaps. Velocity is, after all, the ratio of distance to time and so ordinarily expressed as a number, and not a vector. But the velocity vector delivers the requisite number with an iridescent shimmer. The true velocity or speed of a particle is the *magnitude* of its velocity vector, the vector's twin-chambered heart now generously satisfying intuition.

Just the same series of adjustments and subtle perspectival shifts encompass acceleration as well as velocity. With those adjustments and shifts complete, the calculus has undergone transmogrification, every old concept fully pupated.

❀   ❀   ❀

NEWTON'S *PRINCIPIA* expresses a magnificent scheme for the explanation of change in the universe. It is by no means a simple scheme and so it is very helpful to imagine change appearing in the universe in stages.

Apart from the empty universe, a universe in which there are numbers and functions comprises the simplest of all possible structures. No time beats in this world; there is no change; and nothing happens or has happened or will happen.

Time and space are needed to make these severe dark lines bend themselves into a living world. This Newton brought into existence by means of the laws of absolute time and space.

The continuous functions now provide a record of change in place with respect to change in time. It is the red of movement and so of change that the continuous functions impart, leaving in space a record of where they have been and where they are going.

If the continuous functions lend color to the sketch, they lend depth as well, and this by means of their interpretation as the trace in space of a moving particle. This is an interpretation made grand by its scope, the ordinary world of physical objects (such as that space ship) giving way to a single class of idealized objects and then giving way again in favor of the continuous functions.

So long as the continuous functions leave a geometrical record of themselves as straight lines, and *only* as straight lines, the question where they are going or where they have been can be explained entirely in mathematical terms. There remains the question *why* they are going where they are going. It is a question that within Newtonian mechanics remains unanswerable. Particles traveling in a straight line travel in a straight line because it is their nature to do so. This is the content—and the limitation—of Newton's law of inertia.

Curves are now added to the scheme, particles traveling in great looping arcs or following shallow parabolas to land some distance from where they took off. Mathematics can still explain where these particles are going or where they have been, but to explain *why* they are going there an appeal to Newton's law of inertia is unavailing. A curve is not a straight line. Force is needed to deform a linear universe. This is the province of Newton's second law. And with force in place, a fully *physical* universe appears, Newton's third law guaranteeing the existence of particles engaged in the useful work of transfering forces from place to place The scheme has transformed itself into a richly detailed canvass.

The coordinate system in which time is crossed against space now gives way to purely a spatial system in two dimensions. It is a space overlaid with the filigree of vectors. And it is the vectors that act as a tool of physical dissection, revealing by their happy notational genius connections in the physical world that formerly required the full powers of Newton's own genius before they could be discerned.

❁   ❁   ❁

RISING FROM the origin of a coordinate system, a space ship, or a particle traces a curve in space. The universal clock ticks on and as it ticks its position vector rises from the origin in a straight line to touch its trajectory at a given point.

And it is here and so now that the mathematical and physical apparatus of the *Principia* begin a miraculous fusion.

A vector is a straight line conveying information about direction and magnitude. The direction of the position vector is toward the particle's trajectory, the vector's tip just touching a point occupied by the particle itself and touching each point the particle has reached in turn. Its magnitude is an absolute measure of the particle's distance from the origin.

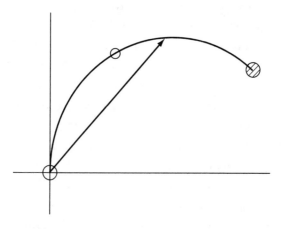

Position Vector Touching the Trajectory of a Particle

It is, that position vector, a straight line in space, and so a geometrical device that can only do what a straight line does, and that is directly to connect two points and so two places. Given a complete record of the position vector as it changes over time, the physicist recovers the ship's curved trajectory. This is tantalizing evidence that vectors are inherently capable of subordinating a world of curves to the unyielding discipline of straight lines.

But the position vector does more and it does more by

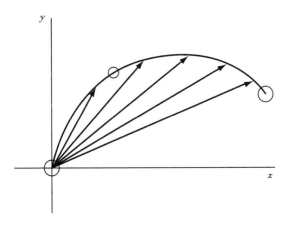

Position Vector Describing a Curve

virtue of the fact that it is inherently capable of dividing its reference. The position vector is a single straight line, an elegant arrow. It records a straight-line change. Its *components*, on the other hand, record positional changes along the x- and the y-axes. The exchange between the vector and its components makes perfect sense. The particle has moved in the x-direction, and it has moved in the y-direction. The position vector is at any point in space the vector sum of its component vectors.

Sums of this sort have a very attractive geometrical interpretation. The position vector corresponds to the diagonal of a parallelogram, one whose outstretched arms mark position along the double axes of the plane.

The resolution of position into its components may be duplicated yet again in terms of velocity and acceleration. The velocity vector is simply a pair of numbers, one that is capable of dividing its reference. Ditto for the acceleration vector.

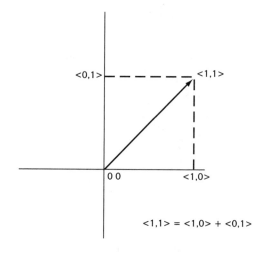

Vector Addition by Components

And if this is something that is seen easily by means of straight lines streaking through the plane, it is something that Newton saw as well, his elaborate geometrical constructions serving precisely the same analytic ends.

The curve of the space ship's (or the particle's) trajectory is now subordinated at every point to analytic domination by vectors and so by straight lines. In the case of the ship's velocity, vector components split the ship's overall velocity into velocity along two spatial dimensions. The ship's on-board speedometer reveals its overall speed; the physicist's internal odometer reveals that the ship is traveling at the speed it is traveling because it is speeding along two quite separate *axes* of a coordinate system. Its overall speed is a fusion of two separate motions, one that may be duplicated at a distance by the separation of a vector into its components and the recombination of those components by vector addition.

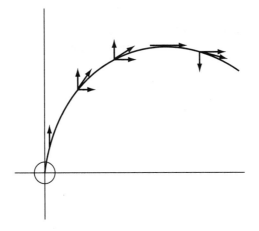

Vector Analysis of Velocity in Terms of Components

Precisely the same division of reference that marks position and velocity marks acceleration as well.

And so, too,—so *thus*—the forces acting upon the ship. It is Newton's second law that forges a connection between force, mass, and acceleration. It is a connection that vector analysis enlarges by treating force itself as a function of time. And with this enlargement in place, mathematical and physical analysis fuse seamlessly.

The forces governing a moving object are the source, and so the explanation, of its change.

And it is now that what is already a richly colored canvass acquires the sheen that only a final layer of paint can provide. The paint is applied by components. One overall force is in action. But like any vector, this one divides its reference, one set of forces directed along one axis; another set, along another axis. Analysis by components distributes some part of the force to one axis, the other part to the other axis, and by the magic of vector addition, resolves those forces into a single vector function. As the product of mass and acceleration, both forces are in effect continuous functions of time, discharging themselves in alignment with the world's ticking clock.

The division of forces now rewards the physicist with a division of labor. The overall force acting on an object acts in *two* dimensions of space. *Component forces act only in one dimension.* This means that component forces may be analyzed in the familiar terms of the calculus in which functions ferry numbers to numbers. For purposes of analysis, vectors disappear. This makes for an immense improvement in understanding. A complicated two-dimensional problem has been resolved into a pair of simpler one-dimensional problems. Thereafter, the original vector force may be recovered by means of a parallelogram of forces.

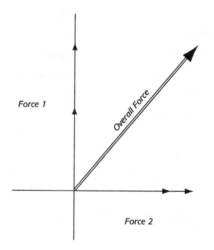

Force 1

Overall Force

Force 2

Parallelogram of Forces

This division of labor ramifies itself downward throughout the analysis, providing a tool for the interpretation of forces, acceleration, velocity, and position.

Analysis by components pries open forces and reveals their internal structure, one force going this way, another that way, the physicist's art now anatomical in the way that it reveals the striated muscles working separately in the smoothly moving limb. It is an art that by its nature reveals certain appearances to be illusions, that caressable limb having an internal structure somewhat at odds with its carressability. In the case of mathematical physics, the evacuation of certain illusions is somewhat more of a surprise. Force brings about a deformation of the world's natural order, causing particles in motion to accelerate and so in some cases to curve themselves in space. The world might seem enriched. Where before there were straight lines, now there are straight lines and curves. But the deformation that force induces, the analysis of force removes. Vectors are straight lines. And straight lines suffice for the analysis of change. The world might seem restored.

As in every artistic effect, it is difficult to say what is the illusion and what the reality. This is hardly a surprise. And it is very marvelous.

❧ 10 ☙

THE ORB OF THE MOON

THE MYTHIC APPLE THAT ONCE FELL IN Newton's orchard has fallen as well in wind-swept orchards far from Woolsthorpe, its echo reverberating and reverberating and reverberating. We hear those echoes; we are persuaded that we quite understand their significance. As a young man, Newton conceived the idea of extending gravity to the "orb of the moon." *Yes, yes,* we are inclined impatiently to say. The earth tugs on the moon, the moon tugs right back. It may have once been very daring to imagine that one and the same force gives both earthbound apples and the moon itself a good solid yank, but gravity, like energy, is now well known for being well known. It requires but two questions to suggest the inadequacy of the common conviction that we understand what Newton meant. Just why must it be *gravity* that is tugging at the moon if something is tugging at the moon at all? This is one question. Given those reciprocating tugs between the earth and the moon, why is it that the moon does not descend directly toward the earth, landing on Iowa City or Tokyo with a monstrous crash? This is a second question. It has not been my experience that a great many people are prepared to answer these questions with confidence. Newton answered both questions so completely that we now take his analysis for granted. This is another measure of his stature.

❁   ❁   ❁

TWENTY YEARS before he completed the *Principia*, Newton described the moon's orbit in terms of a balance of forces, gravitational forces pulling the moon downward and centrifugal forces impelling it outward. This analysis is deficient because it is not general. Within the *Principia*, it is the law of inertia that comes to control the behavior of the moon. Centrifugal forces are not needed; indeed, they are now destined to appear in Newtonian mechanics as fictions.

The law of inertia is entirely a general statement. Throughout the universe, it is straight lines that are natural. Everything else is an artifact of force. Whatever the moon is doing, it is doing because it has been persuaded to deviate from the straight line it would otherwise follow. Sailing through the night, the moon bends continuously around the earth and it bends continuously forever, moving in a circle because the earth's gravitational field makes it impossible to move in a straight line, and moving without stopping because it has displaced its momentum from one path to another.

This analysis is simpler than the one offered by pre-Copernican astronomers. Only one class of geometrical objects—the straight lines—are marked as natural. It is both simpler and more general than Newton's original analysis of lunar motion as a balance of forces—simpler because a balance of forces requires two forces and the analysis by means of inertia, only one. It is more general because it provides, or suggests, a scheme of explanation that goes well beyond circular motion to accommodate *any* physical situation in which force brings about change.

One of the glories of Newtonian mechanics is just that its theoretical claims immediately reveal themselves in vivid and compelling images. Each diagram has a secret that it yields and it is here more than any other place in the *Principia* that it is possible to persuade oneself that one is somehow recapturing Newton's own thoughts as he brought the heavens under the control of his imagination.

Newton's analysis thus begins with the law of inertia. The moon's natural trajectory in the sky is a straight line.

Traveling in a straight line, the moon travels with a certain fixed velocity. It is fixed because no forces are promoting its acceleration.

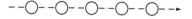

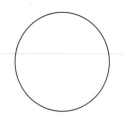

The Earth and the Natural Trajectory of the Moon

And now an observational fact. The moon travels around the earth in a circle.

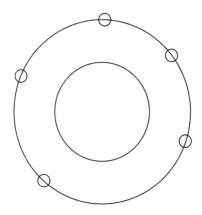

The Circular Orbit of the Moon

An immediate inference follows from Newton's first and second law. If the orbit of the moon is not straight, some force *must* continuously be deforming its path.

But if the orbit of the moon is circular, the force deforming its orbit must be centripetal. It has its origin at the center of the earth. This conclusion follows from Newton's general analysis of circular motion. *Any* object traveling in a circle around a fixed point is controlled by a centripetal force. (In an ordinary carousel, for example, the iron strut connecting the moving horse to the carousel's center provides the force holding fast the horse. Absent the strut, the horse would follow its natural trajectory.)

Newton now invoked his second law for a second time.

Force is needed to deform the orbit of the moon; but force reveals itself in nature as acceleration. It thus follows that at every given moment, *the moon is accelerating toward the center of the earth.*

It is this conclusion that is apt to prompt a puzzle. If the moon is accelerating *toward* the center of the earth, why doesn't it crash *into* the earth?

The answer is revealed by another diagram. The moon falls continuously between the point on its natural trajectory it would otherwise have reached and the point on its orbital trajectory that it does reach.

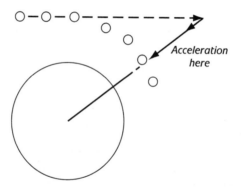

The Acceleration of the Moon*

It falls thus at every moment in time and so it falls forever without ever falling to earth.

The inferences continue to collect themselves. The moon's

*The diagram is, of course, not drawn to scale. and in particular the moon's acceleration is at right angle to its trajectory. See pp. 195-202.

acceleration represents a measurable physical parameter. Using Kepler's laws and the assumption that the moon behaves with respect to the earth as a planet to the sun, Newton next demonstrated that the moon's acceleration is governed by an inverse square force. The inference is not complicated and it is entirely natural.

There now follows the last magnificent inferential step. Given an analytic appraisal of the force deforming the moon's trajectory, Newton compared the magnitude of that force to the force of gravity as it acted on the surface of the earth. He *discovered* that the centripetal pull on the moon, when discounted by distance, and the force of gravity on terrestrial objects, had precisely the same magnitude; he thus *concluded* that they represented one and the same force.

In one grand analytic act, Newton provided an analysis of the moon's motion around the earth and a confirmation of the thesis that he had conceived so many years before, that gravity might be extended to "the orb of the moon."

⇥ 11 ⇤

THE SYSTEM OF
THE WORLD

T HE MOON AND THE PLANETS NOW RE-
cede. A brief retrospective is in order. Newton's analysis of motion succeeded in
brilliantly abbreviating the rich and detailed body of astronomical and theoretical data accumulated over two centuries
by Tycho Brahe, Nicolas Copernicus, and Galileo Galilei. If
nothing else, this would have represented a stunning achievement in pure thought, Newton demonstrating that the very
abstract principles embodied in his laws of motion held
latent the power to control the movement of the moon in the
night sky and the celestial orbits of the planets. But the principles that Newton invoked did far more. They led to a cas-
cading series of immensely productive definitions. The old
physical concept of *work* may, for example, be defined in
terms of the Newtonian concepts of force and distance. With
its tensed mandibles fixing a kernel of corn, an ant does work

to push the kernel forward; force has been applied and distance covered. But if work can be defined in Newtonian terms, energy can be defined as well, and with these definitions in place, the conservation of energy for the first time becomes accessible as a principle of theoretical physics. These definitions have now spread and enlarged themselves so that they cover and explain virtually every aspect of material behavior that is larger than the atom and smaller than the universe.

If on the one hand, Newton's laws take the physicist further and further into theoretical elaboration, on the other hand, they return the engineer to the study of simple and useful mechanical devices. Newton's own remarks are matchless in their concision:

> The power and use of machines consist only in this, that by diminishing the velocity we may augment the force, and the contrary; from whence, in all sorts of proper machines, we have the solution of this problem: *To move a given weight with a given power,* or with a given force to overcome any other given resistance. For if machines are so contrived that the velocities of the agent and resistant are inversely as their forces, the agent will just sustain the resistant, but with a greater disparity of velocity will overcome it.

This is remarkable, the more so since Newton then goes on to remark rather brusquely that "to treat of mechanics [i.e., machines] is not my present business." Of course not, I might add. Machines are not Newton's present business simply because everything that is interesting about machines follows from his real business, which is the analysis of force and mass, and the concepts needed properly to analyze *them*.

❋   ❋   ❋

WHATEVER THE theoretical or practical consequences Newton drew from his laws of motion, his most daring assertion about the universe proceeded from a source of intuition inaccessible to logical analysis alone. His account has the force of some majestic narration. "In the preceding books" [of the *Principia*], he begins:

> I have laid down the principles of philosophy, principles not philosophical but mathematical: such, namely, as we may build our reasoning upon in philosophical matters. These principles are the laws and conditions of certain motions, and powers or forces, which chiefly have respect to philosophy; but lest they should have appeared of themselves dry and barren, I have illustrated them here and there with some philosophical scholiums, giving an account of such things as are of more general nature, and which philosophy seems chiefly to be founded on. . . .

"It remains," Newton goes on to affirm, "that from the same principles, I now demonstrate the Frame of the System of the World." Many of Newton's biographers have drawn attention to this sentence and observed that written by any other figure it would have seemed presumptuous.

It is the law of universal gravitation that Newton saw as the frame of the universe, and what the law says is quite simple. Every material object attracts every other material object with a force that is proportional to its mass and inversely proportional to the square of the distance between them. The mathematical expression has the familiar quality of a sentence spoken in an imperfectly understood language:

$$F = \frac{Gm_1m_2}{r^2}$$

This is Newton's law of gravity. F designates the force between two objects, of course, m_1 and m_2 their masses, r the distance between them, and G is a universal number because it is a universal constant, the same everywhere and for everything.

The law of gravitation constitutes the frame of the universe because its sphere of application is every material object, whether larger or small, here or there, to the uttermost ends of the cosmos. Zooming toward the edge of space and time, a particle no larger than a pebble nonetheless exerts a palpable force against the earth, material objects in all parts of space binding one another and being bound by a connection whose nature is unfathomable in all respects except the mathematical.

Newton's affirmation of the universal law of gravitation recapitulates the chief arguments of the *Principia* itself, and so comprises a natural progression. The barely glittering core of his argument lies far behind, in the *Principia's* first book. It is there that he succeeded in analyzing both lunar and planetary motion in terms of centripetal attraction. In Book III, Newton recalls the development of his own argument. Thus **Proposition I:**

That the forces by which the circumjovial planets are continually drawn off from rectilinear motions, and retained in their proper orbits, tend to Jupiter's centre; and are inversely as the squares of the distances of the places of those planets from the centre.

The same analysis governs **Proposition II:**

That the forces by which the primary planets are continually drawn off from rectilinear motions, and retained in their proper orbits, tend to the Sun; and are inversely as the squares of the distances of the places of those planets from the Sun's centre.

And the same analysis again, **Proposition III:**

That the force by which the Moon is retained in its orbit tends to the Earth; and is inversely as the square of its place from the Earth's centre.

The moons of Jupiter, the primary planets, the moon itself—these all receive a common interpretation in terms of centripetal attraction toward a central source.

Proposition IV is rather different, Newton now affirming

That the Moon gravitates toward the Earth, and by the force of gravity is continually drawn off from a rectilinear motion, and retained in its orbit.

The force of gravity to which Newton appeals is, to be sure, a centripetal force; but it has now been identified with the force governing objects on the surface of the earth itself, Newton deliberately passing from "the Moon gravitates toward the

Earth," which describes only its direction, to "the force of gravity," which describes the nature of the force controlling its direction.

It is within the compass of **Proposition IV** that Newton moves close to the center of his meditations. His own experimental and theoretical analysis had already indicated that whatever the force drawing the moon downward, it is experimentally equal to the force controlling the behavior of ordinary objects on the surface of the earth. In a beautifully controlled thought experiment, he goes on to argue that what his experiments had revealed must indeed be the case.

For imagine, he suggests, that gravity is one force, and that the moon is attracted by quite another force, one that just happens experimentally to have the same power; suppose next that several moons were to revolve about the Earth. One of them—these suppositions are now piling up—is small enough so that it just sails over the highest mountain. The same thought experiment now brings its trajectory to a sudden halt.

What then? Newton asks.

"It would," Newton responds with an air of common sense entirely at odds with the real depth of his discussion, "descend to the Earth, and that with the same velocity with which heavy bodies actually fall from the tops of those very mountains, because of the equality of those forces which oblige them to descend."

This is so very plausible an image—a small moon falling together with, say, a large boulder, both of them falling at precisely the same rate of speed—that for a moment its conclusion lies hidden.

But only for a moment. "And if the force by which that lowest moon would descend were different from gravity, and if that moon were to gravitate towards the Earth, as we find terrestrial bodies do upon the tops of mountains, it would then descend with twice the velocity, as being impelled by both these forces conspiring together."

Experiment and pure thought have in this dream-like sequence fused perfectly, experiment revealing that forces tugging at the moon have the same magnitude as the force of gravity, and pure thought revealing that if this is so they must in fact be identical.

Within a few terse sentences, Newton has expressed the deepest principle of Newtonian mechanics—the identity of gravitational and inertial mass.

❉ ❉ ❉

THE PROGRESSION of four propositions has led Newton to the concept of gravitational attraction between the earth and the moon; in **Proposition V,** the concept is enlarged by a drumbeat of extensions:

> *That the circumjovial planets gravitate toward Jupiter; and the circumsaturnal toward Saturn; the circumsolar toward the Sun; and by the forces of their gravity are drawn off from rectilinear motions, and retained in curvilinear orbits.*

The significance of this drumbeat is too overwhelming to be trusted to a single reading; in his scholium or commentary on **Proposition V,** Newton writes:

*The force which retains the celestial bodies in their orbits
has been hitherto called centripetal force; but it being now
made plain that it can be no other than a gravitating force,
we shall hereafter call it gravity.**

❊ ❊ ❊

THE PROPOSITIONS I have quoted express reason-
ing of surpassing subtlety. Centripetal forces of old have gone
and gone for good. Instead of an inward force attracting the
moon to the earth, or the earth to the sun, there is a reciprocating force, one proportional to the masses of the earth and
the moon, and the earth and the sun. Newton's first law of
motion created a universe in which objects were either at rest
or moving in straight lines. Such was their natural inclination
and force was required to induce their acceleration. Newton's
second law forged an identity between force, mass, and acceleration. The force required to accelerate an object is proportional to its inertial mass. The greater the inertial mass, the
greater the force required to change an object's behavior. It is
an object's *inertial mass* that expresses its obdurate resistance
to change, and it is just because an object has inertial mass—
a kind of inherent sluggishness—that force is needed to
change its behavior in the first place.

The forces needed to express Newton's second law are entirely abstract; not so the force that figures in the universal law
of gravitation. This force has its source in matter itself and

*In coming to appreciate the architecture of the *Principia*, I have profited from S.
Chandrasekhar's *Newton's Principia* (Oxford: 1995) Would that this superb book had
an index.

acts as the supreme expression of the tendency of material objects to draw close to one another. It is an object's *gravitational* mass that expresses an object's innate power of attraction. And gravitational force is, like inertial force, proportional to an object's mass.

Two quite different concepts have entered the discussion—resistance to change and powers of attraction.

Or so it would seem.

Appearances are, in fact, misleading. Gravitational and inertial mass are identical. An object's innate powers of attraction are the same as its innate powers of resistance; desire is a measure of reluctance, just as one might expect on other grounds.

❀   ❀   ❀

WITH PRELIMINARY propositions asserted and then ratified, Newton presents the crucial proposition of the *Principia*—**Proposition VII:**

> *That there is a power of gravity pertaining to all bodies, proportional to the several quantity of matter which they contain.*

The justification that follows is measured, lucid, and triumphant:

> That all the planets gravitate toward one another, we have proved before; as well as that the force of gravity toward every one of them, considered apart, is inversely as the square of the distance of the places from the center of the planets. And thence [by propositions already

cited] it follows, that the gravity tending towards all the planets is proportional to the matter which they contain.

Moreover, since all the parts of any planet A gravitate toward any other planet B; and the gravity of every part is to the gravity of the whole as the matter of the part is to the matter of the whole; and (by Law III) to every action corresponds an equal reaction; therefore the planet B will, on the other hand, gravitate toward all the parts of the planet A; and its gravity towards any one part will be to the gravity towards the whole as the matter of the part to the matter of the whole.

With this proposition, western science passed at once from its inception to full maturity. For the first and only time in the *Principia,* Newton concluded **Proposition VII** with the letters QED—*quod erat demonstrandum.* Thus is it demonstrated.

❆   ❆   ❆

THE CONCEPT of gravity that Newton introduced in the *Principia* is both fruitful and mysterious and so serves as a perfect metaphor for the very activity of rational inquiry. Gravity, although weak at great distances, is the only force that stretches from one end of creation to another. But however weak its force, gravity, since it plays on a cosmological scale, has time in which to play, its powers proceeding by accretion as the universal clock strikes away the seconds, the minutes, the hours, and then the eons. Gravity is thus the only force capable of explaining the large-scale features of the universe. In a universe without gravity, matter would simply disperse itself in space. The universe in which we find ourselves is one in which matter is organized into profound spatial patterns. Objects such as the sun and the earth cohere chiefly because gravity prevents their parts from separating. Beyond the solar system, star systems betray the influence of an intrinsic consolidating force, their shape a reflection—and often a precise reflection—of their mutual attraction. The great galaxies, in which billions of stars are grouped together in the night sky like animals, reveal themselves by their shape to be caught in the same gravitational toils, matter, wherever it appears in the universe, moving toward matter, impelled by an ancient and irresistible force.

If gravity explains much that might otherwise be perplexing, it is an explanation that proceeds by means of a mystery. Gravity acts at a distance and it acts at once. No other force in nature would seem to behave in this way. A force with such properties is hardly mechanical, however much it may be transmitted by material bodies. What is perhaps as perplexing is gravity's irreducible character. There is within Newtonian mechanics no explanation for gravity's force in terms of other forces—the movement or distribution of particles, say. Gravity is what it is; and it cannot be explained in simpler terms or by an appeal to the more elementary constituents of matter. We are acquainted with gravity through its effects; we understand gravity by means of its mathematical form. Beyond this, we understand nothing.

This was true when Newton wrote; it is true today.

❧ 12 ❧

THE CAPTIVE OF HIS CAMOUFLAGE

THE PUBLICATION OF THE *PRINCIPIA* IN 1687 made Newton's reputation. He became well known and then famous. The book was understood at once to be a masterpiece. It was closely read by mathematicians and physicists, and within a short time read by philosophers and men of letters as well. Those who like John Locke accepted its conclusions found in Newton an ally, and those who like Gottfried Leibniz rejected its premises, an antagonist, but whether Newton appeared as an ally or an antagonist he was in either case now a presence too massive to be ignored.

Publication of the *Principia* was by no means an easy task. When Dr. Halley presented the manuscript to the Royal Society, Robert Hooke wasted no time in accusing Newton of plagiarism, seeing in Newton's discussion of the inverse square law simply the elaboration of ideas that he had long conceived but never properly developed. His deep sense of injury was not entirely unjustified; Hooke *had* glimpsed what Newton saw. Although Hooke was immodest, vain, boastful, and tactless, he was not a fool; and had Newton responded to his complaints with good-natured generosity, he would have enhanced Hooke's reputation while doing nothing to diminish his own. Halley pleaded with Newton on Hooke's behalf, reminding Newton that Hooke was ill. He could have as well asked the sea to shed its salt. Newton was incapable of intellectual generosity. The forces that allowed him to create compelled him to destroy. Withdrawing to his study in sullen fury, he struck Hooke's name from the *Principia* entirely. Hooke was never to forget the insult, and Newton was never to forgive Hooke. When seventeen years later, Hooke died, Newton assumed the presidency of the Royal Society that Hooke had vacated. He at once ordered Hooke's portrait to be destroyed.

And so it was.

❁ ❁ ❁

ALTHOUGH THE publication of the *Principia* in July of 1687 would make Newton famous, it had not yet in February of the same year made him politically invulnerable. James II ruled in England and he ruled as a Catholic. Ordering the university to confer a Master of Arts on a Franciscan monk named Alban Francis, the King managed in one splendidly inept gesture to inflame the university's religious sensibilities while offering few emollients to ease its collective conscience. The university protested and then rebelled, otherwise timid academics finding in Newton a champion as unlikely as he was imprudent. Newton kept the gravamen of his religious concerns to himself, of course; there is a considerable difference between anti-clericalism and heresy. Having kept his religious beliefs secret for so long, Newton kept them secret for the duration; he was thus able to court danger without courting catastrophe. Men had in the last decades of the seventeenth century been hanged for far less than Arianism. The Glorious Revolution of 1688 justified his intransigence without revealing its source, the accession of the Stuarts and the departure of James II assigning to England's Catholic minority a voice that carried conviction but not influence.

As a result of his fearlessness, Newton found himself promoted to Parliament. It was an honor both well meant and misplaced. A man incapable of compromise under the best of circumstances, Newton could hardly have been expected to compromise under the worst, and a democratic forum, one whose members were at once pompous and cunning, must have presented itself to his cold eye as a studied exercise in frivolity. There is no record during his tenure at Parliament, except for a request that a window be closed, that Newton ever said anything worth saying. What Newton *did* take from his political experiences was an acquaintance with London and the life that it afforded. The contrast with Cambridge could not have been more striking. London was a metropolis, the largest city in Europe; and Cambridge was yet a village, where at night sheep and pigs roamed the streets untended and dogs barked mournfully at clouds passing beneath the moon. Although brutal, London was also gay; and although filthy, exciting. A simple social system governed Cambridge, the village, with its artisans, shopkeepers, and tradesmen juxtaposed against the university, with its unwilling students and indolent dons. The entire panorama of social life was on display in London—tradesmen and artisans, shopkeepers purveying the world's ornate or elaborate treasures, butchers, beggars, confidence men, thieves, swindlers, prostitutes, stout merchants and their pregnant wives, rich bankers, financiers, shipbuilders, guild members, delicate pretty women, great swaggering aristocrats, courtesans peeping out from behind red lace curtains, politicians, book dealers and publishers, clever witty writers, poets, artists, musicians, and men with skills ranging from leather tanning to drypoint etching.

Newton had finished his great work; he had become a great man.

And great men generally do not live in villages.

Newton's reputation gave him entrance into a society more cosmopolitan than the one he had known. He made the acquaintance of the philosopher John Locke, the architect Christopher Wren, and the diarist Samuel Pepys, who in moments snatched from dalliances had managed to become the president of the Royal Society. Accustomed to deferral, the great lords and politicians took satisfaction in deferring to Newton, thus reinforcing their sense that as deference was owed him by virtue of his genius, so deference was owed them by virtue of their position. On one occasion, Newton was invited to dine with the King, who no doubt managed to suggest acquaintance with the *Principia* while having avoided the inconvenience of studying it closely.

If unfamiliar, the adulation that Newton received was not unwelcome. Men averse to flattery on principle are often unable to recognize it in practice. Newton accepted the great world's admiration and for the first time in his life took pleasure in the company of younger scientists and mathematicians, members of a generation for whom he had become a legend.

Finding himself in an environment that a more cosmopolitan man would have regarded with misgivings, Newton imprudently allowed his heart to swell. The object of his emotional attachment was a Swiss mathematician, Fatio de Duillier. Too young to be a rival and too ingenuous to be a threat, de Duillier had precisely the attributes needed to become a disaster. He was attractive, worshipful, and intellectually impetuous; and when he immodestly suggested that he might improve the *Principia* by rewriting it, Newton regarded his affectation with an indulgence that on any other occasion he would have discharged as indignation. If Newton was taken with de Duillier, de Duillier was taken as well with Newton. For all intents and purposes, Newton was still a man in his prime, his hair, although white, full and luxuriant, and his features finely chiseled. And beyond his physical presence, he was Newton. The men grew close and then mutually dependent, Newton suggesting that de Duillier move to Cambridge and de Duillier suggesting that Newton move to London. There is no way, of course, in which to assess the nature of their relationship beyond recounting its intensity. Their letters reveal ardor, but betray no hint of physical passion. Speculation about seventeenth-century homoeroticism is as irrelevant as it is inescapable. We have no way of knowing what took place between the two men and so must forever alternate between assigning to Newton and de Duillier an intimacy that they did not exhibit or an innocence that they did not enjoy.

It hardly matters.

Newton's relationship with de Duillier appeared to have

renewed his intellectual energies; he worked for two years with his accustomed fury, but the work that he undertook was feverish rather than focused, the fires of his alchemical furnace dancing synchronously with the conflagration in his veins. When his dog Diamond reduced his labors to ashes by knocking over a candle, Newton, understandably unwilling to appreciate the fine symbolism of the accident, was driven almost to despair.

The thing that was waiting to arrive arrived some time in 1692. De Duillier announced his intention to return to Switzerland. Whatever his ostensible reasons—illness and family responsibilities—he had quite evidently found it necessary to end an emotional attachment that could neither be confined nor consummated. De Duillier suffered at giving pain, and Newton at receiving it, neither man notably improved by suffering of the sort that builds character only in those that it does not afflict. And then quite abruptly, their correspondence came to an end.

A deep and intransigent emotional disturbance took hold of Newton's intellect. The crisis announced itself by sleeplessness. Four or five months in Newton's life seem to have vanished, Newton plainly suffering, but suffering in silence. A man never disposed to confronting his emotions could hardly have been expected to confess them. Two painfully poignant letters, written when the most acute stage of his illness had passed, reveal Newton's misery and frustration. "I am extremely troubled," Newton wrote to Samuel Pepys, "by the embroilment I am in." He did not, he added pathetically, "have my former consistency of mind."

The embroilment to which he refers was plainly sexual and just as plainly imaginary. Writing three days later to John

Locke, an embarrassed and contrite Newton begged Locke's forgiveness "for being of the opinion that you endeavored to embroil me with women [and for saying] 'twere better you were dead.'"

Thus was the most powerful scientific intelligence in Europe reduced.

✂ 13 ✂

MASTER OF THE MINT

ISAAC NEWTON COMPLETED HIS SECOND MAS-
terpiece, *The Optics*, during the last decade of the
seventeenth century. Unlike the *Principia*, this
was frankly a work of summation and recapitulation, drawing
in all respects on research that Newton had concluded twenty
years earlier. Written in English, and written in a far more
colloquial style than the *Principia*, *The Optics* brilliantly sum-
marized Newton's counterintuitive theories about light, explaining his conclusions in detail and outlining the exquisite
experiments that had persuaded him his theories were true.

The publication of *The Optics* in 1704 brought Newton's
scientific career to an end. He would thereafter address himself to various trifling problems, but he would never again re-
gain the monumental power and concentrated force needed to
initiate and complete a major endeavor.

It is a measure of Newton's stature that many scientists,
impressed by his achievements, have asked what Newton
might not have accomplished had he continued to work.

There is in this question a note both of awe and disappointment. Newton believed in a corpuscular theory of light; but he was sensitive enough to realize that light also displays wave-like properties. He was unable to reconcile what seemed at the time to be a profound natural contradiction. His biographers have sometimes suggested that had Newton embraced the contradictions that he recognized but rejected, he might have leapfrogged the years and discovered the elements of quantum theory. Perhaps so.

Newton certainly retained the power to contribute to pure mathematics until well past middle age; but he was essentially a physicist and not a mathematician, and he might have made further contributions to pure mathematics only if he might have been a different man.

Newton gave the scientific community its first great physical theory; mathematicians and physicists required more than two hundred years fully to explore the system's strengths and limitations. Newtonian mechanics is now complete. Almost all problems that can be posed within the structures that it provides have been solved. It is only turbulence that remains a significant and baffling question. Had Newton wished, he could well have continued to explore his own creation, setting and solving problems, expanding the margins of understanding, doing the essential work of civilization. He did, in fact, occupy himself with the very difficult three-body problem that arises when the moon, the sun, and the earth are treated as a single complex system. His head aching, he withdrew in frustration from the difficulties he encountered.

But, really, why would Newton have bothered much with the development of Newtonian mechanics? His was a great, an urgent, creative force; and however much he may have been psychologically innocent, Newton at least knew enough about himself to understand that the work that he might now undertake would never quite be work that he would find worth doing.

Newton brought his scientific career to a close for a simple, inescapable and compelling reason: he had nothing more to say.

He would continue to work as he had always worked, but he now worked at other things and he worked to keep busy.

The fires of creation were allowed to damp themselves and then die down.

✻   ✻   ✻

CHRISTMAS DAY, 1696. As the universal clock marked the lowering of the sun with its solemn and awful chime, Isaac Newton reached his fifty-fourth year. He had run a long race and he had run it brilliantly. By the standards of the seventeenth century—by *any* standard, I suppose—he stood at the very threshold of old age. And yet he was in vibrant good health and temperamentally incapable of doing nothing, his nervous system whippet-like in its combination of excitability and rapidly discharging energy. His restlessness conveyed itself to his friends and then to politicians who were prepared to be his patrons; mindful of Newton's stature, Charles Montague—the future Lord Halifax—succeeded in having Newton appointed as Warden of the Mint. Newton made his farewells to Cambridge and to the man he used to be.

Newton's position at the Mint was intended, of course, to be a sinecure. The post, Montague remarked dryly in his letter to Newton, "has not too much bus'ness to require more attendance than you may spare." Montague hardly expected that Newton would work at his job, and other politicians could not demand what Montague did not expect. The arrangement was perfectly ordinary.

Nonetheless, Newton acquired a new role in life at the very moment that English currency was encountering a crisis. The century had rounded itself. Civil war, regicide, and monarchical realignment were wounds that had already formed historical scabs. Even though the war with France continued to sputter and then flare up, a rich and vibrant commercial culture was coming into existence.

The great trading houses and the government had a variety of financial instruments at their disposal, of course; but the general currency of English life was based on metal and stamped in coins. It was a currency that counterfeiting and clipping had steadily degraded. Every government regards counterfeiting with alarm. By diluting the quantity of coins in circulation, counterfeiting has the effect of debasing the currency as a whole. But if counterfeiting is a threat to the state, it is also a temptation to the criminal simply because it is as easy to accomplish as it is hard to detect. Seventeenth-century England was awash in counterfeiters and as a result, English coinage was looked on with suspicion abroad and contempt at home.

Facing a monetary crisis, Parliament accepted the obvious. English currency would have to be completely recoined. At the same time, Parliament passed legislation fixing counterfeiting as high treason, punishable by death. If drastic measures were needed in order to protect the stability of English currency, politicians with the most to lose from its debasement were perfectly prepared to do away with those with the most to gain.

It fell to Isaac Newton to undertake both tasks, Newton discovering in himself administrative skills that surprised even his warmest political supporters. Even though he was not yet Master of the Mint, the position still occupied by the time-serving Thomas Neale, he quickly took command of the Mint's machinery and by means of superb organization and an incredible attention to detail, succeeded in streamlining the Mint's operation. Men who had been accustomed to snoozing away their sinecures suddenly found themselves staring into Newton's demanding eyes, or faced with the prospect of answering his punctilious memoranda. The Mint's machinery was overhauled; extra horses were pressed into service; and production placed on virtually a twenty-four-hour basis, the Treasury ordering the Mint to open its doors at four in the morning and close them again at midnight. Newton did not simply accept authority: he embraced it naturally, his letters to his superiors a model of cool efficiency and adroitly calculated servility. Other great scientists have discovered in themselves a hidden talent for administration. It was Robert Oppenheimer, after all, who managed the cranky and contentious mathematicians and physicists responsible for the development of the atomic bomb. Newton's case was different. The Mint was not a scientific institution, and recoinage had nothing to do with mathematical physics.

Newton's reinvention of himself as a civil servant represents an unusual personal achievement. He had never been the most organized of men, his scientific work undertaken in long but manic bursts. He had no doubt been utterly and completely meticulous in his experiments, and fantastically

mindful of detail, but he had been long accustomed to work
that had been undertaken entirely for its own sake, and while
his intellectual efforts had been models of great clarity, no
one, impartially surveying the numerous projects that he left
unfinished could honestly conclude that Newton's chief talent
was a gift for bureaucratic organization.

And yet apparently it was. A man who for thirty years had
worked as he pleased, oblivious to the constraints of time and
indifferent to demands that he publish his results or commu-
nicate his ideas, now found himself the manager of a complicated financial institution. He was expected to be punctual
and orderly, responsive to his superiors and solicitous of his
workmen, accessible when required and authoritarian when
needed. Whatever the expectations, Newton met them. Nothing so suggests the power of his personality as this remarkable
transformation. It was a transformation that apparently Newton accepted as ordinary, evidence again that this most remarkable of men lacked only the capacity, or the desire, to
analyze himself.

❋   ❋   ❋

IF NEWTON WAS a successful administrator, he was
an equally successful prosecutor, the full weight of his malignant fury finding prompt and easy expression in the campaign
he launched against counterfeiters. Seventeenth-century English law was harsh, but seventeenth-century English justice
was by no means arbitrary. The authority of the Crown was
open to legal challenge, and those with the money or influence
to acquire attorneys often found the principles of English law
a source of personal satisfaction. If the standards of proof re-

quired in court were often appallingly loose, suspects at least faced conviction in a court where standards of proof applied. Since few counterfeiters chose to operate their presses in plain view, counterfeiting was an especially difficult crime to detect, and since counterfeiting was widespread, a man apprehended with counterfeit coins could justifiably observe that he came by those coins just because counterfeiting had proven so difficult to detect.

After an initial period of uncertainty, Newton set himself to break the wheel of counterfeiting with his customary energy and what in retrospect seems a singular ruthlessness. He left his offices in the Mint late at night to make the rounds of low taverns and grog shops, even petitioning the Treasury for special disguises so that like Sherlock Holmes he might pass unnoticed in shady haunts. With a fine disregard for conflict of interest, he had himself appointed justice of the peace in nineteen counties. His criminal interrogations were said to be especially fearsome, something that Newton himself must have understood, since he ordered all records of those interrogations destroyed. One can only imagine the circumstances. The men Newton was determined to destroy were almost without exception petty criminals. Counterfeiting is, after all, a distributed crime, one whose effects are cumulatively but not individually grievous. Unlike theft, counterfeiting harms no one directly, and men willing to break the law in the first place could hardly be prepared to appreciate the fact that when multiplied, individual peccadilloes might well mount to a general catastrophe. Newton thus faced men prepared to deny their guilt because they had antecedently dismissed the significance of their crime. He quickly overcame their scruples. And what makes the convocation between Newton and

his adversaries so bizarre is its simple injustice. Facing the
most powerful intelligence in Europe, and facing that intelligence without counsel, how on earth could a common criminal propose either to dissemble or remain silent when, under
English law, he had every right to do both?

To be sure, Newton, since he possessed little psychological
acuity, could hardly have wasted his time on psychological
stratagems. His tactics were simpler and they were more effective. He made himself master of detail; his tavern spies
having roamed the London gutters for months, Newton was
in a position to confront suspects with an elaborate record of
overheard conversations, whispered betrayals, and dangerous
secrets. Faced with Newton's extraordinary ability to command the details of their life, many petty criminals agreed to
act as his informers, Newton ultimately succeeding in his
prosecutions simply because at the appropriate time, he had
accumulated men prepared to say what he wished them to say.

His most celebrated prosecution was conducted against
William Chaloner, rather a more exotic rogue than he habitually encountered. Chaloner was a man who had managed to
promote himself fluidly through the complex class structures of English society. At the time he faced Newton, he was
wealthy and he was influential; but he had acquired his
wealth by illicit means and the men who were prepared to
befriend him were also prepared to betray him. Chaloner
was an inspired counterfeiter, the perfection of his work
alarming both Newton and the Treasury. He had as well a
certain perverse flair for playing one organ of government
against the other, and when the occasion arose to obtain a reward by falsely accusing entirely innocent printers of print-
ing Jacobite propaganda, he saw his opportunities and he

took them. The printers were hanged, Chaloner rewarded. Utterly underestimating the man that he now faced, Chaloner conspired to throw a shadow of doubt over the Mint itself, suggesting that the Mint might profit from his own coining device—after all, who knows coins better than a counterfeiter?—and then recklessly exacerbating his own impertinence by hinting to a government commission that the Mint was itself involved in various shady coining operations.

Newton was moved to fury. Over the course of two years, he made himself the master of Chaloner's life, accumulating witnesses when possible and suborning them when necessary, and like prosecutors everywhere, disposed to test the web of his evidence by its complexity rather than its truth. Chaloner was both wicked and imprudent. It is not clear that he was guilty. He was nonetheless convicted of high treason and sentenced to death. His attorneys were powerless to save him, and his friends indifferent to his fate.

Having exhausted his appeals, Chaloner determined to beg for pity. "Most merciful Sir," he wrote to Newton, "I am going to be murdered although perhaps you may think not but tis true I shall be murdered the worst of all possible murders that is in the face of Justice unless I am rescued by your merciful hands." The letter is signed "your near murdered humble Servant." Chaloner's letter, needless to say, if it attracted Newton's attention failed to elicit his compassion. On the 23rd of March, 1699, Chaloner was hanged, drawn, and quartered.

If Newton discerned in his own behavior tendencies incompatible with English law, it is not something he ever disclosed.

Few states have ever commanded men of Newton's intelligence, but every state has commanded men of Newton's temperament.

And, of course, they still do.

⌐ 14 ⌐

THE DEFILE

SAAC NEWTON WAS APPOINTED MASTER OF
the English Mint in February of 1700, succeeding the odious Thomas Neale who had, by sheer
force of perseverance, managed to cling to a position that he
had done nothing to deserve and very little to enhance. Newton would remain the Mint's Master until his death. Just three
years later, Robert Hooke brought his side of his controversy
with Newton to a close by succumbing to the afflictions that
for more than seventeen years had made him wretched. Newton was elected president of the Royal Society; it was a position commensurate with his stature and with his influence. At
the age of sixty-one, he had completed the transformation of
his personality and acquired a powerful institutional identity.
A man who in his youth and middle age had been diffident
and defensive now moved with assurance through the corridors of political and intellectual power.

During his years at the Mint, Newton retained at least some
of his matchless mathematical powers. On New Year's Day 1697,
Jacob Bernoulli issued a challenge "to the sharpest mathemati-

cians in the whole world." Given two points A and B in the vertical plane, Bernoulli asked, what is the line of quickest descent a particle would follow, supposing that A and B are joined by a thin wire? That wire is intended only to guide a particle in its trajectory, and the particle is moving downward under the force of gravity. Such is the problem of the *Brachistrochrone*, as it has come to be known. The political circumstances that prompted the challenge are not entirely clear, but if Bernoulli had hoped for an answer from Continental mathematicians, he must have been disappointed. The problem lay unanswered for a month. Having heard nothing from the men most likely to have discovered its solution, Bernoulli forwarded his problem to Charles Montagu, the president of the Royal Society. It passed from Montagu to Newton. Newton's niece, Catherine Barton, has recorded Newton's reaction. On January 29, 1697, she wrote,

> Newton did not come home till 4 [in the afternoon] from the Tower there to find awaiting him a printed paper bearing a twin mathematical challenge from a young Groningen professor, Johann Bernoulli, addressed generally "to the sharpest mathematicians flourishing throughout the world." Newton did not hesitate straightaway to attack the problem, indeed, he did not sleep till he had solved it which was by 4 in the morning.

Newton sent his response to Montagu in a letter dated January 30, 1697; it was then published anonymously in the *Philosophical Transactions* for January of the same year.*

*In presenting this anecdote, I have followed S. Chandrasekhar's *Newton's Principia for the Common Reader*, pp. 571–572.

Newton's solution of the *Brachistrochrone* is a model of cool efficiency. "From the given point A," he writes,

> Draw the unbounded straight line APCZ parallel to the horizontal and upon this same line describe both any cycloid AQP whatever, meeting the straight line AB (drawn and, if need be, extended) in the point Q, and then another cycloid ADC whose base and height shall be to the previous one's base and height [as AC:AP] respectively as AB to AQ. The most recent cycloid will then pass through the point B and be the curve in which a heavy body shall, under the force of its own weight, most swiftly reach the point B from the point A. *As was to be found.*

A cycloid, I should add, is the curve traced by a circle (or a wheel) as it moves horizontally along a straight line.

Although Newton published his solution anonymously, Bernoulli knew at once that only one man might have written it. "Although its author," Bernoulli remarked, " . . . does not reveal his name, we can be certain beyond any doubt that the author is the celebrated Mr. Newton: for even if we had no information other than this sample, we should have recognized him by his style even as the lion by its paw."

❀　　❀　　❀

THE CIRCUMSTANCES of Newton's life now lose their power to fascinate. He lived as other men in his position might live; he worked assiduously, kept regular hours, he attended to his distant family, and if he indulged himself neither at his table nor at any of the fashionable London fleshpots, this was only because abstemiousness had made him indifferent to food and unresponsive to women. By all accounts, his life in London was sober, unavailing, diligent, and boring.

His massive intellectual weight impressed itself both on the institutions that he dominated and the future that he controlled. As president of the Royal Society, Newton imposed his personality on what had until then been a somewhat chaotic organization, determining the problems it would address, its agenda, and its style. His influence on the Royal Society and on English scientific life was not altogether benign. His genius lacked for only one quality: it could not be copied. Rational mechanics underwent development in the eighteenth century, but its leading ideas were modified by Continental mathematicians—Bernoulli, Laplace, Lagrange. But Newton gave to English scientific life the luster of his own accomplishment, but what might have been a rich stream of English mathematics dried up before it could even begin to flow.

Newton had never in his life been disposed to accept criticism; as Master of the Mint and president of the Royal Society, he was now in a position to dismiss what he did not wish to hear and to deny what he could not dismiss. His rage remained as ferocious as ever, and when he *did* find his will checked, his indignation was, I suppose, the more intense because provoked by circumstances that had become unexpected. Nothing so inflames a man grown accustomed to deference as its denial. The great, the monumental, controversy with Gottfried Leibniz etched itself acidly on his soul. Both men enlisted their numerous allies in their own defense and conducted themselves as if the issue of who had first dis-

covered mathematical ideas that both had conceived independently was something that might be settled by invective or
insinuation. This may suggest peevishness in men whose
achievements should have made them indifferent to issues of
priority; and, no doubt, both Newton and Leibniz had allowed
their pride to color their judgment. But peevishness suggests
as well that Newton and Leibniz were hurling abuse at one
another only to establish just who could lay claim to the discovery of the calculus. More was at stake. Newton had made
the calculus his own matchless but idiosyncratic instrument;
Leibniz had given his system of analysis a notation of uncommon elegance and flexibility. Both men must have sensed
that the very future of their creation was somehow associated
with their name and hence with the question of priority. And
both men were right. Newton fused the calculus very closely
to mathematical physics and so thoroughly covered his steps
that no mathematician was able to follow in his path. Leibniz
opened the calculus to the learned world. And the learned
world gratefully followed his lead.

The controversy between the two men continued until
Leibniz's death in 1717; and thereafter, although Newton continued to fume, the audience that he had hoped to impress
turned its attention elsewhere and sailed calmly on.

❀   ❀   ❀

NEWTON WAS never a romantic man, but he had a
passionate nature, and facing old age and the decline of his
scientific powers, he discovered in himself a certain current of
warm sympathy, a kind of alien tenderness. He was by 1696
reasonably well off and took care to provide for his mother's

side of his family, purchasing an annuity for his mother's sister that quite literally rescued her and her children from poverty. His niece Catherine Barton came to live with him in his house on Jermyn Street. It was rather an unusual arrangement. Catherine was a young woman; Newton already an old man. What is more, Catherine was by every account a woman of striking beauty and uncommon wit. Newton was plainly taken by his niece, but there is nothing whatsoever to suggest that if his senses had been stirred his better judgment had been corrupted. He had reached the age in which even very great men are capable of appreciating beauty without wishing to possess it.

If Newton was content to regard his niece with paternal affection, not so Lord Halifax. Letters leave no doubt that Newton's patron fell tempestuously in love with Newton's niece, thus introducing into Newton's household a free-floating current of illicit erotic energy. Great love affairs are often their own justification, and to his credit, Newton did nothing to destroy what he could otherwise do nothing to discourage. He would, I imagine, leave for work and return, dine and retire, all the while taking a curious satisfaction as a lacework of passion accumulated around him. The domestic arrangement that Newton tolerated inevitably became the subject of conversation. Voltaire had been Newton's warmest French admirer, but he was an incorrigible gossip and when he visited London in the 1720s, he determined to his endless satisfaction that Lord Halifax had secured Catherine Barton's affections by first handsomely rewarding her uncle. Voltaire wasted no time in reporting as fact what he could only have learned as rumor. "I thought in my youth," he remarked, "that Newton made his fortune by his merit." Evidently not:

Isaac Newton had a very charming niece, Madame Conduitt, who made a conquest of the minister Halifax.
Fluxions and gravitation would have been of no use
without a pretty niece.

If Lord Halifax had initially been prepared to bribe his
way into Catherine Barton's bedchamber, he was certainly not
averse to remaining there by the same means. His will of
1706 made generous provision for the young woman, and the
revisions in the will made what had been generous extravagant. "These Gifts and Legacies," Halifax wrote, "I leave to
her as a Token of the sincere Love, Affection and Esteem I
have long had for her Person, and as a small Recompense for
the Pleasure and Happiness I have had in her Conversation."

When the astronomer John Flamsteed learned of these
remarks, he wrote to a friend that Halifax's bequest was made
to Catherine Barton for "her *excellent conversation*." Newton
had for years hounded poor Flamsteed for astronomical data,
finally going so far as to order Halley actually to seize and
publish his astronomical books. Flamsteed's *fine* observation
rounds out a little Restoration drama, one involving youth,
beauty, hopeless infatuation, spitefulness, and benign maturity.

It was a drama that did no one harm and almost everyone
credit.

❋   ❋   ❋

THERE IS VERY little more. In 1717, Catherine Barton married John Conduitt, a man ten years her junior, and a
commissary to the British naval forces on Gibraltar. The cou-

ple provided Newton with a home, and while neither of them made any pretensions of understanding his mathematical work, both quite understood the magnitude of his achievements. Conduitt diligently recorded a great many details of Newton's life, his record as unrelieved as the life itself must have been uninteresting. Newton continued to work as he had always worked. He allowed the focus of his attention to return to Biblical studies; he wrote and endlessly rewrote a *Chronology of Ancient Kingdoms*. He lived the life he had chosen for himself. And he remained completely the captain of his camouflage, the secrets of his soul concealed from all eyes, his own included.

When at last he came to face death, he chose to describe himself in words of majestic detachment:

> I don't know what I may seem to the world, but as to myself, I seem to have been only like a boy playing on the sea shore, and diverting myself in now and then in finding a smoother pebble or a prettier shell than ordinary, whilst the great ocean of truth lay all undiscovered before me.

⊰ 15 ⊱

THE QUEST

NEWTON STANDS ALONE. THERE IS NO underestimating his influence; there is no evading his authority. In creating the science of rational mechanics, Newton determined the goals of mathematical physics as well as its methods. A subject that had been a tale became a quest. Physicists now think of physics in terms of ever more general and thus ever more universal laws, seeing in gravitation, electromagnetism, relativity and quantum mechanics steps in a progression converging towards principles so powerful that they will provide a complete account of matter from its smallest to its largest state and from the beginning of creation to its very end.

There have been in the tide of time four absolutely fundamental physical theories: Newtonian mechanics, of course, Clerk Maxwell's theory of electromagnetism, Einstein's theory of relativity, and quantum mechanics. They stand in history like the staring stone statues on Easter Island, blank-eyed and monumental. Each is embedded in a continuous mathematical representation of the world; each succeeds in amalgamating far-flung processes and properties into a single, remarkably compressed affirmation, a tight intellectual knot. The supreme expression of each theory is a single mathematical law, one expressed as an equation, a statement in which something that is unknown is specified by hints and whispers, contingencies arranged in a certain way. And each of the great theories contains far more than it states, the laws of nature fantastically compressed, as if they were quite literally messages from a timeless intellect.

The quest has now achieved a degree of accuracy that must be reckoned miraculous. Quantum electrodynamics commands the very heart of matter to something like twenty decimal places. In determining the distance from New York to Los Angeles, theory and measurement diverge by no more than the width of a single human hair. The great theories are singular. No other intellectual effort commands their prophetic powers, pure thought and physical experience coinciding to a degree that is never achieved in any other domain of intellectual or practical life, coinciding, that is, to a degree that is unprecedented in the entire experience of the race, a specification of points and places in the future utterly at odds with our habitual inability to say where our lives may have been misplaced or our hearts irretrievably lost.

Mathematical physicists are now persuaded that the Newtonian quest is coming to an end because its goal is coming into view. What Newton sensed, they sense they can now see.

❋   ❋   ❋

WHATEVER THE benefits of the scientific age, the consolations of religion are not commonly thought among them. No matter Newton's own religious convictions, the displacement of faith into irrelevance has been widely attributed to Newton's quest. "The universe," Carl Sagan has argued, "is everything there was, or is, or will be." This claim, although widely made, is rarely defended, perhaps because it seems a logical truth, since its negation is a logical contradiction. If transcendence denotes the desire to transcend *everything*, it is incoherent as a doctrine and useless as a conviction. The contrast that Sagan intended to draw is between a world of matter and a world beyond matter. It is a world beyond the world of matter whose existence Sagan was prepared to deny. Sagan is hardly alone. Some form of materialism has become a contemporary orthodoxy. Scientists disposed to see the universe as nothing less than a physical system are quite naturally prepared to see the universe as nothing more than a physical system. Theirs has not been a view that has elicited widespread enthusiasm even though it has commanded widespread assent. If Newton's quest has seemed magnificent, it has also seemed monstrous. Human beings, it would seem, have a remarkable capacity for intellectual ingratitude.

The laws of physics occupy a singular place in intellectual life. They are the point to which all cones must taper. This reductive metaphor derives its meaning from another. If nature is organized as a book, it is a book organized in turn as a deductive system, one grander than the *Principia*, to be sure, but no different in kind. Within that book—or *the* Book, to assign the metaphor its pride of place—every statement about the material world finds its explanation in deductive terms. If the planets describe a conic circle, this is because all objects are bound to one another by the universal law of gravitation, the law standing to the fact as the premises of an argument to its conclusion. With the *Book* read, the universe will be known. Having already fallen into disfavor, transcendental categories of thought will fall into disuse.

All of this represents, to be sure, a collection of metaphors converging toward the future. Nonetheless, metaphors have a strange unexpected life of their own, and often make demands that they cannot entirely satisfy. The attempt to explain the biological world in terms of the laws of physics has not been a notable success. "We cannot see," Richard Feynman wrote in his remarkable lectures on physics, "whether Schrödinger's equation [the fundamental law of quantum theory] contains frogs, musical composers or morality."

If we cannot *see* that the laws of physics specify the properties of living systems, then plainly we do not *know* whether the Book of Nature is complete.

It is fair to observe that this objection, since it depends on knowledge that we do not possess, makes claims that we cannot assess. It is an issue that only time will resolve. The metaphor of nature organized as a deductive book runs into trouble in the here and now. If nature is organized as a book, just what explains its premises? They are, those premises, descriptions of the material world; they provide an account of its most general properties. If the Book of Nature is complete, they, too, will require an explanation. An appeal to still further laws is, of course, ruled out on the grounds that the Book that results would either be infinite or circular.

An appeal to logic is unavailing. The laws of nature are not logical truths.

So, too, an appeal to simplicity. The laws of nature must be intrinsically rich enough to specify the panorama of organic life and life is anything but simple. "Blind metaphysical necessity," as Newton remarks, "which is certainly the same always and everywhere, could produce no variety of things."

It would seem to follow either that the laws of nature are not statements about the material world or that the Book of Nature is incomplete. But if they are not statements about the material world, what, then, are they statements *about?* And if the Book of Nature is incomplete, what reason is there to reject transcendental categories of thought?

Some three hundred years after the publication of the *Principia*, Newton's quest has encountered a strange bifurcation, the great vector of mathematical physics either directing itself beyond the material world or hurtling forward along an endless path. Newton was himself a religious man; he had no doubts where the vector would turn. In drawing the *Principia* to a close, he expressed his most general theological views in rolling prose. "This most beautiful system of the sun, planets and comets, could only proceed from the counsel and domination of an intelligent and powerful Being." The theological specifics that follow place God everywhere and at every time. "He is eternal and infinite," Newton writes, "omnipotent and omniscient; that is, his duration reaches from eternity to eternity; his presence from infinity to infinity; he governs all things and knows all things that can be done." Although the Deity exists *always* and *everywhere*, his existence is not corporeal. "He is utterly void of all body and bodily figure, and can therefore neither be seen, nor heard, nor touched. . . . We have ideas of His attributes, but what the real substance of anything is, we know not. We know Him only by his most wise and excellent contrivance of things."

A great many of these theological claims are already implicit in the scientific portions of the *Principia* itself. The universal law of gravitation is the mathematical description of a force whose existence is known by its effects. Any attempt to lift the veil of appearance reveals yet another veil, the world's "secret springs" forever hiding themselves. "Hitherto," Newton writes, "we have explained the phenomenon of the Heavens and of our sea by the power of gravity, but we have not yet assigned a cause of the power." The material world cannot completely explain the principles by which it is most successfully described; the system that we see can "only proceed from the counsel and domination of an intelligent and powerful Being," one "very well skilled in mechanics and geometry."

When this statement is divided by the concept of a powerful being, what is left is the concept of intelligence itself. Since they are the most general statements about the physical world, the laws of physics comprise the supreme record of intelligence as it exercises itself in matter. To the extent that the world's intelligence cannot be fully explained in terms of the world's matter, the laws of physics are transcendent.

Intelligence is, of course, a term that we understand in anthropomorphic terms and Newton quite plainly conceived of the supreme being as if he possessed an intellect rather like his own. The role that intelligence plays in Newton's discussion—it is surely not an argument—has been played variously by such concepts as complexity, organization, and even beauty. These are concepts that in the final analysis are as hard to avoid as they are difficult to explain. It might be the better part of wisdom to replace the more glamorous transcendental categories with a single, hard-working concept. The laws of physics, Newton might have observed, are not *ar-*

bitrary. They seem to describe a very particular organization of the universe, one possibility among many. If we could come to understand that most possibilities for the organization of the universe were more or less likely, then we could come to understand the laws of physics by an appeal to the laws of chance. This would extend Newton's quest but not complete it, for inevitably we would wish to know why the initial probability distribution that governs the laws of physics had the shape that it did and took the form that it took. This question again transcends the ambit of material objects.

It is no wonder that some physicists and many philosophers, when asked for an explanation of the laws of physics themselves, are content to say simply that every chain of explanations must come to an end. This response, however gratifying, plainly leaves unanswered questions that continue to torment the human imagination. It is altogether too easy to affirm that explanations come to an end precisely when they are most needed.

It may seem an act of ingratitude to suggest that Newton's magnificent vision of the physical sciences has ended in an inconclusive quest. Still, if we are troubled and intellectually confused, we are troubled and intellectually confused in ways immeasurably more sophisticated than were possible before Newton tortured himself into thought.

The goal remains far away, but we have as a species collectively covered a very great distance.

APPENDIX

DESCENT INTO DETAIL

A Brief Mathematical Chrestomathy

A *function*—the distance function $D(t) = t^2$, for example—specifies a relationship. The values of $D(t)$ for $t = 1, 2, 3$, and 4 are 1, 4, 9, and 16, and so on and so upwards. Various different functions specify various different mathematical relationships.

The *average speed* of a particle moving under the control of a distance function is, over an interval of time, the ratio of change in distance to change in time.

The *instantaneous speed* of a particle moving under the control of a distance function is, at a given time, the limit of a series of average speeds.

The *average acceleration* of a particle moving under the control of a velocity function is, over an interval of time, the ratio of change in speed to change in time.

The *instantaneous acceleration* of a particle moving under the control of a velocity function is, at a given time, the limit of a series of average accelerations.

Ordinary English now passes into ordinary symbols.
Difference in time is:

$$\Delta t;$$

and difference in distance

$$\Delta D.$$

The *contraction of time* is

$$\Delta t \to 0$$

Average speed is

$$\frac{\Delta D}{\Delta t}.$$

Instantaneous speed as $\Delta t \to 0$ is

$$\lim \frac{\Delta D}{\Delta t}.$$

The *derivative of D(t)* is designated by the symbol

$$\frac{dD(t)}{dt},$$

so that

$$\frac{dD(t)}{dt} = \lim \frac{\Delta D}{\Delta t}$$

as $\Delta t \to 0$. This is a definition.
The derivative of the distance function, when specified for

various times, yields a new function in turn—the velocity function *vel(t)*.

$$\frac{dD(t)}{dt} = \lim \frac{\Delta D}{\Delta t} = vel(t).$$

Average acceleration is

$$\frac{\Delta vel(t)}{\Delta t}.$$

Instantaneous acceleration, as $\Delta t \to 0$, is

$$\lim \frac{\Delta vel(t)}{\Delta t}.$$

The second derivative of $D(t)$ is

$$\frac{dD^2(t)}{dt} = \lim \frac{\Delta vel(t)}{\Delta t}.$$

This, too, is a definition; and this definition also yields a new function in turn—the acceleration function *acc(t)*

$$\frac{dD^2(t)}{dt} = \lim \frac{\Delta vel(t)}{\Delta t} = acc(t).$$

A *differential equation* is an equation in which an unknown function is specified by means of its derivative. For example

$$\frac{dx}{dt} = 2t.$$

The solution of this equation is $x = t^2$.

The calculations needed to make these definitions work are unfamiliar, but rudimentary. A particle has been propelling itself in space. The distance that it covers is under the control of the distance function $D(t) = t^2$. Nothing has changed. Everything is the same. What, then, is its speed at precisely the very moment the clock has struck two?

At $t = 2$, the particle has covered 4 miles; one hour later, at $t = 3$, 9 miles.

Δt is thus 1 hour.

$\Delta D(t)$ is thus 5 miles.

$\Delta D(t)/\Delta t$ is thus 5 miles in one hour.

Δt now contracts.

If $\Delta t = .5$, then $\Delta D(t)/\Delta t$ is thus **4.5** miles in 30 minutes.

If $\Delta t = .25$, then $\Delta D(t)/\Delta t$ is just **4.25** miles in 15 minutes.

If $\Delta t = .10$, then $\Delta D(t)/\Delta t$ is just **4.10** miles in 10 minutes.

The series of numbers **5, 4.5, 4.25, 4.10** . . . is plainly tending toward a limit at the number 4.

Just so. It is the number 4 that is assigned to the tangent line as its slope; and the number 4 that again measure the particle's instantaneous speed at the point $t = 2$.

Units of Measurement

FUNDAMENTAL

Distance	The meter = m
Mass	The kilogram = k
Time	The second = s

APPENDIX

DERIVED

Speed Meters Units: Meters per second (ms⁻¹).

Acceleration Meters Units: Meters per second, per
second $(\text{ms}^{-2} = \frac{\frac{m}{s}}{s} = \frac{m}{s^2} = \text{ms}^{-2})$.

Force Newtons Units: Kilograms times meters
per second, per second (km⁻²).
1N = Force required to accelerate
1 kilogram to one meter per
second, per second.

Two Important Constants

G $6.68 \times 10^{-11}\,\text{Nm}^2\text{kg}^{-2}$

g $9.8\,\text{ms}^{-2}$

The units of the universal gravitation constant, G, are chosen so that the right side of Newton's law of gravity $F = G\frac{M_g m_g}{\rho^2}$ is expressed entirely in terms of Newtons, and so in terms of force. The value of *g* is a measure of gravitational acceleration near the surface of the earth.

The Mathematics of Change

Consider a particle moving along a straight line. Its mass is known. And it is moving as the result of a constant force. Some years ago, the physicist Freeman Dyson imagined a rocket ship propelled through space by the solar winds. It is lovely image and might well serve as the picture behind the example.

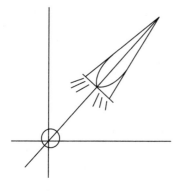

**Rocket Ship Sailing
Through Space**

Moving along a straight line from the origin of a coordinate system, the space ship is conveying itself from place to place within an inertial reference frame. The reference frame serves only to mark distance; the ship's direction is analytically unavailable. This is plainly a miniature and thus a contrived coordinate system, but in mathematical physics, as in so many other things, it is only the contrivances that give us a clue to the complexity beyond.

That rocket ship is now sailing serenely for places that human eyes have never seen. It has long since left the pale blue earth behind, following a straight line past the inner planets of the solar system, and then its outer planets; the trace of its trip has long vanished from optical and radio telescopes. It is voyaging toward infinity. What we have on earth—we who have conceived a hopelessly romantic interest in its strange journey—is an abiding theoretical interest in the distance it has covered as time unwinds from its silken spool. Mathematical physicists recognize that interest as a demand for the space ship's path or trajectory. If we could know how far the silver ship has gone in terms of how long it has gone far, we who have been left behind should know all that we can know.

An appeal to Newton's second law is now before the earthbound court. The ship has long since dwindled to a dot and then disappeared. It now acquires a remote identity as a particle. The force acting upon it is a mystery. The solar wind? Perhaps. Or perhaps something even more exotic, a kind of pumping energy borrowed from the future. Whatever its nature, it appears in all further calculations as a constant. Wherever the ship is going, it is going by means of one and the same force.

Enter now Newton's second law: force is the product of mass and acceleration, or

$$F = ma.$$

These symbols have not penetrated the real world's carapace if only because acceleration has not yet been correlated with time. That correlation is analytically accessible by means of the calculus.

—**Acceleration is a function of time;**

thus on substituting $Acc(t)$ for a in Newton's law

$$F = mAcc(t).$$

Dividing both sides of this equation by m yields

$$F/m = Acc(t).$$

But

—**acceleration is also the derivative of velocity.**

By definition, $dvel(t)/dt = Acc(t)$, and so

$$F/m = dvel(t)/dt.$$

Symbols have been substituted for symbols; Newton's law has been algebraically rearranged. But nothing has yet been accomplished.

Nonetheless, the last revision in Newton's second law prompts what lawyers term a leading question. *What* unknown function, when differentiated, yields F/m? It is the velocity function $vel(t)$, to be sure; so much is clear from the symbols themselves, but what, the question really urges, is its *mathematical* nature? This the symbols do not indicate.

Putting the question in the form of a differential equation:

$$dx/dt = F/m?$$

the variable x functioning here as a stab into the dark.

The specification of x is provided by the calculus.

$$x = \frac{F}{m}t + c_1.$$

With the identity of x revealed, the mathematical nature of velocity emerges into the light. Velocity is determined by three familiar arithmetical operations—division, multiplication, and addition—acting on numbers measuring force, mass, time, and something that for the moment remains obscured beneath the symbol c_1.*

The proof? When differentiated, the function $\frac{F}{m}t + c_1$ deposits F/m.

The space ship's velocity has now been described and described explicitly as a function of time.

---

*The choice of units is, of course, inessential, and miles per hour is homier than meters per second.

APPENDIX

❀   ❀   ❀

Precisely the same elementary techniques may be used to recover the identity of the space ship's *position* as a function of
time. Indeed, the same pattern in which a remembered identity is followed by a leading question quite suffices. If acceleration is the derivative of velocity,

**—velocity is also the derivative of position or
displacement.**

But the identity of the velocity function is known; it has
been revealed. And it is on record. This prompts the drumbeat
of a second question. *What* function, when differentiated,
yields $\frac{F}{m}t + c_1$? It is, to be sure, the distance function $D(t)$, but
what is wanting, as before, is a specification of its mathematical nature.

Putting the question in the form of a second differential
equation:

$$dy/dt = \frac{F}{m}t + c_1?$$

The answer is again provided by the calculus:

$$y = \frac{1}{2}\frac{F}{m}t^2 + c_1 t + c_2.$$

At any given moment, these new symbols say, the displacement of the spaceship is determined by this formula.
Time ticks, and at each tick, the formula does its work of division, multiplication and addition; when the numbers have
been manipulated, the tick of time has been correlated with
the displacement of distance.

The space ship's position has been described and described explicitly as a function of time.

❀   ❀   ❀

Two functions have now emerged on the scene, their identities no longer a mystery.

They are velocity

$$Vel(t) = \frac{F}{m}t + c_1$$

and distance

$$D(t) = \frac{1}{2}\frac{F}{m}t^2 + c_1 t + c_2.$$

The darkness has receded. But there are two constants in these formulas as well; their identities have remained undisclosed. A further series of suppositions is required to endow these symbols with their intended meaning.

—Suppose that no time at all has gone by.

It follows that

$$t = 0$$

Substitution yields

$$Vel(0) = \frac{F}{m}0 + c_1$$

and then

$$Vel(0) = c_1$$

the equation's middle term dropping away when multiplication by zero has done its work.

It follows that the constant c_1 represents the ship's velocity when no time has passed—its *initial* velocity.

By the same token,

$$D(0) = \frac{1}{2}\frac{F}{m}0^2 + c_1 0 + c_2,$$

and thus

$$D(0) = c_2,$$

multiplication by zero again wiping away the intermediate debris between $D(0)$ and c_2. The second constant represents the ship's distance from the origin as the exercise begins—its *initial* position.

Velocity and distance have now been completely specified, each symbol playing an explicit role. The classical instrument of mathematical physics is at hand in the form of equations with specified initial conditions.

❀   ❀   ❀

A remarkable exercise has just been concluded. Time itself has been brought under the control of a symbolic apparatus. The future has been deprived of its mystery. It is now as accessible as the present, at least along the dimensions of speed and distance. The exercise is remarkable in another respect. It reveals, as perhaps only a contrived example can, the precisely integrated way in which mathematical and physical concepts mesh within the scheme of understanding advanced by the *Principia*. Time, space, distance, velocity, acceleration, force, and mass are physical concepts. They attach themselves to a world of particles in motion. But the relationships among these concepts is mathematical. Some relationships involve little more than the elementary arithmetical operations; others, the machinery of the calculus. Without these specifically mathematical relationships, Newton's laws would remain unrevealing.

The *Principia* introduced into human thought the novel idea that formulas can be engaged to control the physical behavior of a moving object. This idea was current before Newton, of course, but in comparison to the abridgement of experience provided by Kepler and Galileo's laws, Newton's laws are incomparably richer in their content and they control a far more generous range of observation. They are richer in a second sense as well. Kepler and Galileo provided a description in mathematical form of certain physical facts. The explanations they provided are dynamical. They bring about a correlation of parts between change in place and change in time. But they do not provide an explanation. By assimilating physical change to change in speed, and by correlating change in speed to force, the *Principia* makes possible what physicists call a *kinematic* interpretation of the facts. It is an interpretation that in the *Principia* becomes an explanation, the first entirely serious scientific explanation in human history.

The explanation is partial, and it is incomplete. This Newton surely knew. There is much in nature that cannot be explained in terms of purely mechanical forces—electricity and magnetism, to take one example, the behavior of light, to take another. But inasmuch as the *Principia* deals only with mechanical forces, its success is almost absolute.

Within the Newtonian universe, the laws governing the behavior of particles in motion hold dominion over the past, the present, and the future. They are in this sense deterministic; this is remarkable again. Our appreciation of the natural world is profoundly rooted in our own human sensibility and whatever we may say in response to various philosophical doctrines, no aspect of that sensibility is more secure than our conviction that we bring the future into being by a free exercise of our will. Newton's discovery that very significant aspects of the physical world behave in ways that would appear to have nothing to do with the exercise of *any* will is, when properly understood, deeply disconcerting, the more so since the same pattern of differential specification that controls the contrived example I have given also controls the fall of the moon in the night sky and the revolution of the planets around the sun.

The Earth and the Moon

Newtonian physicists must, in order to justify the analytical claims that they make, forge a threefold link between the straight line that represents the moon's natural trajectory, the curve that it obviously follows in the real world, and the forces acting to change the moon's orbit *from* a straight line *to* a curve.

The analysis proceeds by means of a double impulse. The first is purely mathematical. No forces are involved, and little physics—nothing beyond motion, in fact. As always, a coordinate system. Only two dimensions are required. The moon's orbit is almost planar. The earth is at the center of the system and the moon travels around the earth in more or less a circle. These facts may be expressed entirely in terms of obliging straight arrows. These arrows function as vectors and so allow the physicist to bring the full weight of analysis to bear on the physical situation.

A radius vector **r** is an arrow going from the earth to the moon. Linking two points, it functions as a mark of position.

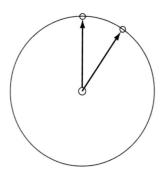

The Radius Vector

Expressed as a function **r**(*t*) of time, the radius vector, like the minute hand of some great clock, sweeps counterclockwise around the circle as time beats on. At any given moment, the function **r**(*t*) determines a pair of numbers lying on the circumference of the circle.

The familiar geometry of the circle is now brought into play, yielding an analytic expression for **r**(*t*):

$$\mathbf{r}(t) = \text{r}\cos(\omega t)\mathbf{i} + \text{r}\sin(\omega t)\mathbf{j}$$

The sine and cosine are in this expression what they always are—trigonometric expressions; time is simply time; and the symbol ω denotes the angular velocity of the radius as it moves around the circle, the speed with which this minute hand is moving. Unit vectors are unit vectors. This formula reveals the position of the radius vector on the circle's circumference. (Thus suppose the radius and angular velocity are 1; suppose further that the time is 0. The radius vector will touch the circumference of the circle at the point cos(0) + sin(0). As it happens, cos(0) = 1 and sin(0) = 0. Thus $\mathbf{r}(t) = <1, 0>$.)

This may well seem an elaborate way of drawing a circle in space. *Seem?* It *is* an elaborate way of drawing a circle in space; the notational complexity is justified by the astonishing wealth of information it provides.

Given the radius vector $\mathbf{r}(t)$, velocity and acceleration vectors $\mathbf{v}(t)$ and $\mathbf{a}(t)$ are free for the asking: $\mathbf{v}(t) = d\mathbf{r}/dt$ and $\mathbf{a}(t) = dv/dt$.

Four conclusions now follow, and they follow at once.*

In the first place, the velocity and radius vectors are always perpendicular; they move at right angles toward one another.

This means that the velocity vector has *no* component along the radius. No part of the particle's velocity is being discharged downward.

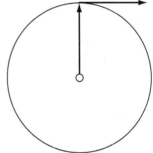

The Velocity and Radius Vectors

In the second place, the velocity and acceleration vectors are always perpendicular; they, too, move at right angles toward one another as both circumnavigate the circle.

---

*These conclusions follow *mathematically* from the simple vectorial description of the moon's orbit. No other facts are required. Details are available in any textbook—J .M Knudsen and R. J. Horth's *Elements of Newtonian Mechanics* (Springer-Verlag, Berlin, Germany, 1995), for example.

APPENDIX

The velocity vector lies along a straight line tangent to the curve; the acceleration vector thus points inward, toward the center of the system. It is useless, of course, to imagine the moon accelerating downward along the full shaft of the acceleration vector. If it did, it would crash into the earth. Although the acceleration vector is pointing downward, the moon's acceleration is discharged in its movement *from* its inertial line *to* the surface of the curve.

In the third place, the acceleration vector has a simple analytic expression in the formula

$$1 \qquad\qquad \mathbf{a}(t) = -\frac{4\pi^2}{T^2}\mathbf{r}(t).$$

The letter T designates the moon's orbital period—the time it takes to complete a single revolution. The formula indicates that the acceleration is negative, the particle picking up steam in the direction of the earth.

And finally, the particle's speed—an ordinary number, now—and its acceleration—another ordinary number—can be expressed by two very simple formulas:

$$2 \qquad\qquad v = r\omega$$

and

$$3 \qquad\qquad a = \frac{v^2}{r}.$$

These numbers correspond to the magnitude of their respective vectors.

These five facts have very straightforward English equivalents.

—There is a formula for the trajectory of a particle
   moving around the earth in a circle;
—the particle's velocity and its acceleration vectors
   move at right angles to one another;
—ditto its velocity and position vectors;
—there is a formula for the particle's speed around the
   circle,
—and another for its acceleration toward the circle's
   center.

❀   ❀   ❀

Enter now the multiplication of *musts*. If the moon travels
around the earth in uniform circular motion, it neither speeds
up nor slows down. Its velocity is thus the ratio of its circumference ($2\pi r$ of high school memory) divided by the period of
its revolution, or distance divided by time

4 $$v = \frac{2\pi r}{T}.$$

Velocity figures in both equations 3 and 4; getting rid of it
entirely (by substituting $\frac{2\pi r}{T}$ for v in equation 3) yields

5 $$a = \frac{4\pi^2 r}{T^2}.$$

This is a formula expressing the magnitude of the moon's acceleration in terms of just four numbers.

❀   ❀   ❀

Thus far mathematics and a little movement along a curve. Enter now physics. Kepler's laws are laws of planetary motion. The moon is not a planet. Still, there is no reason why Kepler's laws should not apply to the system comprised of the earth and the moon, the system amounting to a local solar system, something convenient and close at hand.

Now Kepler's third law affirms that the time taken by any given planet to revolve about the sun, when squared, is proportional to its average distance from the sun, when cubed. In symbols

$$6 \qquad \frac{r^3}{T^2} = C,$$

where C is a constant contingent on the mass of the sun and so always the same for all of the planets.

But as it happens

$$7 \qquad \frac{4\pi^2 r}{T^2} = \frac{4\pi^2 C}{r^2},$$

Since

$$\frac{4\pi^2 C}{r^2} = \frac{4\pi^2}{r^2} \times \frac{r^3}{T^2} = \frac{4\pi^2 r}{T^2}.$$

There is thus only

$$8 \qquad a = \frac{4\pi^2 C}{r^2}.$$

This is an equation of great power, indicating as it does that the acceleration of the moon toward the center of the earth is *inversely proportional to the square of its distance from the earth.* For the special case of circular motion, Kepler's laws of planetary motion have just released Newton's inverse square law of gravitational attraction.

Enter now force by means of Newton's second law:

$$8 \qquad\qquad F = ma.$$

But equation 7 provides an analytic expression for acceleration, and so

$$9 \qquad\qquad F = m\frac{4\pi^2 C}{r^2}.$$

Full stop. And thence a return to earth. Equation 9 is simply a record of something or other tugging at the moon. What is needed in all this is some connection between that tug and the force of gravity on the earth.

Like any other force, terrestrial gravitation obeys Newton's second law:

$$10 \qquad\qquad F = mg,$$

where g measures the acceleration due to gravity.

If that lunar tug and terrestrial gravitation are one and the same force, it follows that it should be possible to substitute $\frac{4\pi^2 C}{r^2}$ (from equation 8) for g:

$$11 \qquad\qquad g = \frac{4\pi^2 C^*}{\rho^2},$$

where C^* is a constant now contingent on the mass of the earth, and not the sun, and where ρ^2 is the radius of the earth, and not the radius of a planetary orbit.

We can, in fact, do more. Since from equation 6, we know that

$$12 \qquad C^* = \frac{r^3}{T^2}.$$

g itself should be

$$13 \qquad g = \frac{4\pi^2 r^3}{\rho^2 T^2}$$

when $\frac{r^3}{T^2}$ is substituted for C^* in equation 11.

But 13 gives the physicist, and it gave Isaac Newton, an expression in which every particular item can be measured. The list is as follow:

T = twenty-seven days, forty-seven hours and three minutes;

r = 60.1 times the radius of the earth itself;

6.37×10^6 meters = the radius of the earth.

When these values are inserted into equation 13, the result is an estimated acceleration of 9.8 meters a second.

The acceleration in question is the acceleration of an object *near* the surface of the earth; but it has been calculated by means of figures derived *from* the acceleration of the moon.

It is possible to measure the acceleration of objects near the surface of the earth using a simple pendulum, Newton himself having carried out just such measurements. The result is again an acceleration of 9.8 meters a second.

This almost perfect coincidence almost completely justifies the claim that one and the same force acts on the moon and on objects near the surface of the earth. A powerful theoretical conclusion has been vindicated by a precise quantitative measurement. Nothing very much more complicated than vectors has been used in the demonstration, that and a little algebra. The coincidence of two numbers and the reasoning by which it has been demonstrated represents a complete triumph of mathematical physics.

The Fall of the Moon

Under ordinary circumstances, we live our lives as terrestrial chauvinists, drawing a clear and apparently irresistible distinction between what takes place on the surface of the earth and the Great Beyond beyond. This distinction is as old as memories of the Garden of Eden; and it has been revived in recent times by the unutterably lovely pictures of the earth taken from space. It is against this weight of common sense that Newton's demonstration must be appreciated. Nothing in our collective experience—*nothing, but nothing*—suggests that the great forces of ordinary life might find room to play in the empty space beyond the earth. And yet they do. This we now know, but it is difficult knowledge and humanity required genius for its acquisition. It is the more astonishing that Newton's specific conclusion about the magnitude of gravity on the earth and on the moon is a part of a more comprehensive system of insights.

One and the same force controls the behavior of falling apples and the moon. Naïve common sense scruples at this conclusion simply because ordinary objects fall toward the center of the earth even as the moon continues to sail serenely throughout space. Newton's qualitative account of the moon's orbit plausibly rejects the conclusions of common sense. It is possible for the moon continuously to fall *toward* the earth without ever falling *to* the earth.

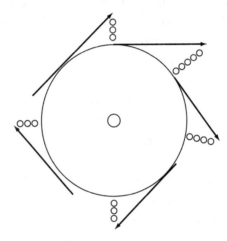

The Moon's Eternal Fall

Precisely the same conclusion, reached qualitatively on p. 133 has been prefigured in the equations running from 1 to 13 on pages 190–194, the secret there all along. Begin then with equation 8, where an inverse square law holds. The magic of mathematical inference works backward in this case, the physicist retracing his steps until he reaches equation 1. It is equation 1 that determines the moon's orbit and equation 1 determines that it *must be a circle.* From the assumption that the earth is attracting the moon by means of an inverse square law, the shape of the moon's orbit follows as an inescapable conclusion. Although the moon falls *toward* the earth, it does not fall *to* the earth because it travels *around* the earth.

Newton's analysis of the moon does not yet amount to the complete analysis that Newton reached in answering Dr. Halley's question about the orbit of the planets. The planets travel in elliptical orbits; the demonstration involves more work. But the reciprocating relationship between equations 1 and 10 do indicate the nature of Newton's grander accomplishment. Setting aside details, the mathematical machinery shows that Kepler's laws entail an inverse law of attraction *and* that an inverse law of attraction entails Kepler's laws. From a mathematical point of view, Kepler's three laws are exhausted by the demonstration that they are equivalent to an inverse square law. By means of a stunning mathematical simplification, Newton forged an extraordinary connection between two disparate physical items—a shape and a force.

The elegance of this conclusion suggested to Newton, as it suggests at once to us, that it is the force of gravity that is crucial; given that force, the facts follow.

The Fall of a Cannon Ball

Standing on the top of Mt. Everest, a physicist imagines that the center of the earth is the origin of a coordinate system. The radius of the earth is ρ. He then shoots a cannon due east, the cannon ball traveling in a direction parallel to the horizontal axis of the earth. Its initial velocity is simply v. It commences to move. As it does, it also commences to fall, its rate of descent toward the center of the earth governed by Galileo's law $y = \frac{1}{2}gt^2$. Two motions, then, in two different directions—out and over and down and out. These motions

APPENDIX

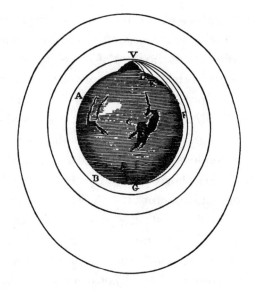

Newton's Illustration from the *Principia*

are independent. No part of the object's horizontal velocity goes down, and no part of its vertical acceleration goes out. Vector analysis makes this plain, the relevant vectors always at right angles to one another and so carrying on without discharging velocity in one direction or acceleration in another.

One second has now passed. The object has traveled $v \times 1$ distance forward, and it has fallen $y = \frac{1}{2}g1^2$. Suppose that x marks the spot the cannon ball would have reached had it not fallen at all. The figure connecting ρ to Mt. Everest, Mt. Everest to x, and x back to the center of the earth via $(\rho + y)$ is a simple right triangle.

A right triangle. It follows from the geometry of the right triangle that

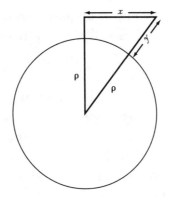

The Cannon Ball's Path

$$\rho^2 + (v \times 1)^2 = (\rho + y)^2.$$

Do the multiplication and let S stand in for $(v \times 1)$; the result is

$$\rho^2 + S^2 = \rho^2 + y^2 + 2\rho y$$

But y^2 is tiny in comparison to $2\rho y$. It follows that

$$S^2 = 2\rho y,$$

at least as a very good approximation.

After a few algebraic adjustments and substitutions, this equation gives way to

$$v = \sqrt{\rho g}.$$

The identity on display connects the cannon ball's initial velocity with something else, a number most obviously. The number is not at issue; it has dropped into the stream of this argument by means of various algebraic twists and turns. It is the nature of that number that is crucial.

If $v = \sqrt{\rho g}$, whatever $\sqrt{\rho g}$ happens to be, then the cannon ball's orbit must be circular. The argument proceeds by accretion. Falling sixteen feet in the first second, the cannon ball has fallen to a point on a curve, roughly at the tip of a radius vector. What holds for the first second, holds for the next second as well, and as time beats on, it holds for all points on the circumference of the circle. Falling to the circumference of a circle at every point, the cannon ball must describe a circle.

But the full weight of numerical secrets stands revealed only when specific values are substituted for various numbers. The radius of the earth is known: $\rho = 6.4 \times 10^6$ meters; ditto, the acceleration due to gravity: $g = 9.6$ meters^{-2}. The cannon ball's initial velocity must therefore be 8×10^3 meters^{-1}.

At this velocity, the cannon ball will trace a circle in space whose radius is roughly equal to the radius of the earth. Let that cannon ball go just a little bit faster, and the radius of the circle it traces in space will be just a little bit longer than the earth's radius.

This cannon ball will fall forever toward the earth without ever falling to the earth.

This is a powerful result, one confirming Newton's thesis that lunar and terrestrial motion are essentially the same by demonstrating that terrestrial objects would behave as the moon does behave if given sufficient initial velocity. Needless to say, what Newton justified by means of his superb visual imagination and his powerful mathematical technique is confirmed every time a satellite is sent into orbit.

There is very little in the history of the exact sciences that combines pictorial vividness and precise quantitative reasoning in an intellectual figure of immense plausibility.

In fact, there is nothing else.

Inertial and Gravitational Mass

Within the Newtonian system, there are two forms of mass, inertial and gravitational. Inertial mass describes that property of a material object that causes it to resist acceleration; gravitational mass, that property that causes it to attract other material objects. Inertial mass is a measure of resistance; gravitational mass, a measure of attraction. Yin and Yang, as Chinese philosophers say (on every conceivable occasion).

Inertial and gravitational mass find their expression in two different parts of the Newtonian system. Newton's second law of motion draws a connection between force, *inertial* mass and acceleration. There is no appeal to gravitation. Newton's universal law of gravitation, on the other hand, draws its connection between force, *gravitational* mass and acceleration. There is no appeal to inertia. Two different conceptions of mass; two different laws of nature:

1
$$F = m_{inertial}a$$

2
$$F = G\frac{M_{gravitational}m_{gravitational}}{\rho^2}$$

In equation 2, G is a universal constant, M, the gravitational mass of one object, and m, the gravitational mass of another. The number ρ measures the distance between these objects.

APPENDIX

A few chiropractic adjustments suffice to change equation 2 into

$$3 \qquad F = m_{gravitational}\left[\frac{GM_{gravitational}}{\rho^2}\right]$$

Near the earth, $\left[\frac{GM_{gravitational}}{\rho^2}\right]$ is known to have the value $g = 9.8$ ms^{-2}, and so functions as a measure of local gravitational acceleration. With another adjustment, equation 3 goes over to

$$4 \qquad\qquad F = m_{gravitational}\,g,$$

and so recovers the form of equation 1.

We now perform two thought experiments. In the first, an object A is placed on a flat surface and attached to a retaining wall by a spring. Only inertial forces are in play, the string's tension measuring the force needed to overcome A's obdurate inclination to stay put. Newton's first law is in play; Equation 1 in charge.

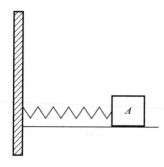

The First Experiment

That spring is now severed and A is rotated—nothing more—so that it is above the surface of the earth. As the experiment begins, A is at rest: it has zero initial velocity. Shortly thereafter, A begins its expected downward acceleration. The requisite force prompting this obvious effect must lie with Newton's first law and thus with Equation 1. After all, nothing in the experimental set-up has changed except for A's spatial position.

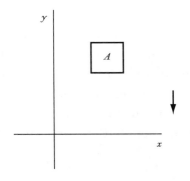

The Second Experiment

But we physicists know as well that the force acting on A can only be the force of gravity. What else is there? Newton's second law of motion and thus Equation 2 provides a second controlling interpretation of the facts. And so we must have

$$5 \qquad F = m_{inertial}\, a = m_{gravitational}\, g,$$

or simply

$$6 \qquad m_{inertial}\, a = m_{gravitational}\, g.$$

Were this not true, Newton's laws would be in conflict.
Dividing through by $m_{inertial}$ we have

$$7 \qquad a = \left[\frac{m_{gravitational}}{m_{inertial}}\right]g$$

An observational fact now enters the discussion.

8 *All objects near the surface of the earth fall at the
same rate of speed.*

From 8, it follows at once that the ratio $\left[\frac{m_{gravitational}}{m_{inertial}}\right]$ must
be equal to 1, and that as a result, $m_{gravitational}$ and $m_{inertial}$ must
be the same. Follows? Follows how? By means of the observation that if inertial and gravitational masses were not the
same, then g would not be a constant, one assuming the same
value for all material objects. If, in turn, g were not a constant,
8 could not be true.

NEWTON CHRONOLOGY

The dates are Julian, with the new year starting on January 1.

1642

> Galileo dies.
>
> Newton's father dies before Newton is born.
>
> Newton is born on Christmas day at Woolsthorpe Manor, Lincolnshire.

1643

> Newton is baptized.

1646

> Newton's mother Hannah remarries Barnabas Smith and moves to Smith's manor.
>
> Newton's grandparents inhabit Woolsthorpe to raise him.

1653

> Newton's stepfather dies. Hannah returns to Woolsthorpe.

1655

> Newton goes to King Edward VI Grammar School in Grantham.

1660

Newton continues at Grantham to prepare for university study.

1661

Newton is admitted to Trinity College, Cambridge.

1662

Newton experiences a religious crisis and composes a "list of sins."

1663

Lucasian chair is established. Isaac Barrow is appointed its first professor.

1664

Newton explores Descartes and outlines the *Questiones Quaedam Philosophicae,* a set of study topics.
Newton is elected to a scholarship at Trinity.

1665

Newton is graduated from Trinity as a bachelor of arts.
Plague hits Britain. Newton returns to Woolsthorpe.
In isolation at Woolsthorpe, Newton begins work that would inspire his later works in mathematics and optics.

1666

Newton expands on the topics outlined in *Questiones.* He formulates his theory of the calculus in bursts of compulsive exploration.
Newton improves Galileo's calculation of the force of gravity by first calculating centrifugal force.

Newton builds on ideas of Hooke, Descartes, and Huygen. Experimenting with prisms results in *Of Colors*.

1667

Newton returns to Trinity and is elected a fellow.

1669

Newton publishes *On Analysis by Infinite Series*.
Newton is elected Lucasian professor of mathematics.
Newton describes in writing the reflective telescope.

1670

Newton's *Lectiones Opticae* explore prismatic phenomena.

1671

Newton begins *A Treatise on the Methods of Series and Fluxiones,* which he never finished.

1672

Newton's reflecting telescope is reviewed by the Royal Society.

Newton sends a written account of his *Theory of Colors* to the Royal Society. (*Theory* is thought to be a preliminary expression of what would become the *Hypothesis of Light,* originally composed in response to Hooke's proclamations that Newton plagiarized his *Micrographia.*)

Newton is elected fellow of the Royal Society.

Newton's interest in mathematics and optics wanes, while his interest in chemistry and theology comes to the fore. He explores the validity of the holy trinity.

1675

Newton's *Hypothesis of Light* is sent to the Royal Society.

Leibniz visits the Royal Society and is given access to
Newton's writings on mathematics.

1676

Newton sends his conciliatory letter to Hooke, including
the "on the shoulders of giants" metaphor.

1677

The Royal Society's Oldenburg dies.

1679

Newton is present at his mother's death. He stays at
Woolsthorpe through the year to set her affairs in order.

Hooke assumes role of secretary of the Royal Society. He
resumes correspondence with Newton on planetary
motion.

1683

John Wickins, Newton's closest friend (his chamber mate at
Trinity), leaves the college.

1684

Halley, Wren, and Hooke attempt to apply the inverse-
square relation to celestial movement.

Edmund Halley consults Newton about elliptical orbits.

1685

James II, a Catholic, ascends the throne.

1687

Newton and other Cambridge delegates appear before Judge Jeffreys's Ecclesiastical Commission in the Alban Francis case. The Francis case is concluded in favor of Cambridge. Newton's *Principia* is published.

1689

Newton is elected to Convention Parliament.
Newton attends a dinner with King William.

1693

Described as Newton's "black year." Newton falls into depression, pursues bureaucratic positions in London.

1696

Newton is appointed Warden of the Mint and moves to London.

1697

Coiner William Chaloner is imprisoned.

1699

Chaloner is convicted of treason and hanged.
Master of the Mint Thomas Neale dies.

1700

Newton is appointed Master of the Mint.

1701

Newton resigns his chair and fellowship at Cambridge.

1702

Queen Anne ascends the throne.

1703

Robert Hooke dies.

Newton is elected president of the Royal Society.

1704

Newton presents *Opticks* to the Royal Society.

1711

The priority dispute begins with Leibniz over the invention of the calculus.

1713

The second edition of *Principia* is published.

Newton appoints Halley as a Royal Society secretary.

1714

Queen Anne dies. George I ascends the throne.

1716

Leibniz dies.

1717

A second English edition of *Opticks* is published.

1721

A third English edition of *Opticks* is published.

Halley retires from the Royal Society.

1726

A third edition of *Principia* is published.

1727

Newton dies and is buried at Westminster Abbey. Newton's writings of that time reveal his rejection of Trinitarianism.

1728

The Chronology of Ancient Kingdoms Amended is published posthumously.

INDEX

INDEX

Index